"Great way to bring new thinking on the innovation process to an organization!"

—*Richard Gendon, Managing Director, Center for Professional Management*

"Praveen has captured a practical way to tap into our latent innovation potential to create winning ideas for sustained competitive advantage in an increasingly dynamic business environment."

—*Mahender Singh, Ph.D., MIT Center for Transportation and Logistics*

"Creating the new can't be the exclusive purview of an isolated team. Praveen Gupta has provided a confluence of ideas and approaches to engage the greater organization in sustained innovation."

—*Bob Aron, Ph.D., Director, New Product Development, DeVry University*

"I am impressed by Praveen Gupta's work in the field of innovation. I believe he has created the science of innovation to help everyone become more innovative."

—*Shanmugham Mahalingam, Ph.D., President and Director, Delhi Business Academy*

"Praveen Gupta has guided us as to how we can become innovative. The most powerful weapon of the future must be innovation, so we can call him a genuine innovation evangelist of this age who will lead the world to change."

—*KyuSan, Cho, Senior Economist, The Bank of Korea*

BUSINESS INNOVATION

In the 21st Century

PRAVEEN GUPTA

Library of Congress Cataloging-in-Publication Data

Gupta, Praveen.
  Business Innovation in the 21st Century
  1. Innovation 2. Business 3. Management
  ISBN 1-4196-4664-8 (Hard Cover)
  ISBN 1-4196-4663-X (Paperback)

Library of Congress Number: 2006931993 (Hard Cover)
Library of Congress Number: 2006931992 (Paperback)
Published by: Accelper Consulting, and
BookSurge, LLC, An Amazon.com Company
North Charleston, South Carolina

Volume printing and binding by Thomson-Shore, Inc., Dexter, MI. United States of America

For obtaining permission to translate in other languages, to buy from stock or for receiving bulk quantity discount please contact Accelper™ at:

Fax: (847) 884 7280
E-mail: info@accelper.com
Tel: (847) 884 1900

Dedicated to

*Dr. Peter F. Drucker for his visionary work
on the discipline of innovation in the 1980s,*

and

*EVERYONE (kids and adults)
for sharing knowledge with me.*

About the Author

Praveen Gupta has been differentiating his work throughout his career. He has solved many problems over 25 years at small to large organizations. Praveen has developed a Six Sigma Business Score-card for measuring corporate performance, the 4P (Prepare, Perform, Perfect, Progress) model for achieving process excellence, the CorporateSigma for monitoring Six Sigma initiative, and Brinnovation™ for accelerating innovation. Praveen, president of Accelper Consulting, works with corporations in achieving sustained profitable growth. He is a frequent speaker at conferences and special events.

Praveen is the author of several books, including *Six Sigma Business Scorecard* (McGraw Hill), and *The Six Sigma Performance Handbook* (McGraw Hill). Praveen holds a BSEE from IIT Roorkee, India, and a MSEE from the Illinois Institute of Technology, Chicago, Illinois. Besides consulting, he teaches Operations Management at DePaul University, and Business Innovation at Illinois Institute of Technology, both in Chicago, Illinois.

CONTENTS

PREFACE

Business Innovation in the 21st Century sounds like an ambitious project. This book was undertaken to evolve the understanding of the innovation process, which is so little understood for so long. At first the task appears to be impossible to accomplish; however, having so many accomplices with a common goal, this project is now a reality. I am proud of my team of authors, editors, reviewers, practitioners, students, professionals, and academicians who shared the same excitement for my framework of innovation as I did. They all have contributed to the book with one purpose in mind: to make innovation a real process that can help society at all levels around the world.

The book took about two and a half years of work, thousands of hours, and in some respects may still appear to be a work in progress. In reality, the book started a journey in which we can all participate to enhance our curiosity and apply our most distinguishing asset: the brain. Thinking is hard, but that capacity is what makes us human. We must practice purposeful thinking to accomplish anything in our professions or to do well in society.

The book presents a comprehensive approach to the innovation process. While writing as well as editing various contributions, I recognized that some aspects appear to be repetitive; however, all aspects are actually presented slightly differently and thus continue to enhance the understanding of this subject. Interestingly, this book accommodates business as well as technical aspects of innovation, which is an unusual combination. This innovative aspect of the book is preserved here to be one source of information in the technology age, where everyone must be somewhat innovatively applying technology and business sense to develop solutions or strategies.

One of the most frequently asked questions regarding this book is, "Where are the case studies?" This book is not about providing some cookie cutter recipe. However, the book does present examples and cases integrated throughout the book. Besides, considering the unique integration of various available and practiced approaches, one can see around the case studies of various principles presented in the book. If you just look, you can find many companies practicing these approaches naturally and sporadically, but not necessarily in a planned and systemic manner. In a nutshell, this book presents a system that I have called Breakthrough Innovation, or Brinnovation.

I trust and hope that the book offers value to every reader, be it the business reader or the technical professional. I would love to hear your feedback, successes, and recommendations.

Praveen Gupta
President, Accelper Consulting
praveen@accelper.com
September 10, 2006

ACKNOWLEDGEMENT

Interestingly the acknowledgement is the final piece in writing a book. However, acknowledgement is a continual thought that facilitates the writing process. Recognizing everyone is impossible, because doing so could create a book in and of itself. Nothing can be done without someone's help, and thus acknowledgement is certainly appropriate. This book, being a challenging task, was supported and helped by numerous people.

First of all, acknowledging the people who helped me start the project is a must. Avanti, my daughter, gave me insight into the innovation process. I would like to thank my colleagues Cissy Pettenon, Rajeev Goel, Rick O'Brien, Mike Lippitz, Rajiv Khanna, Rajeev Jain, Anoop Verma, Tarun Kumar, Glen Nevogt, Baber Inayat, Abhas Kumar, Laurie LaMantia, Rajesh Tyagi, Paul Davis, Marjorie Hook, Hans Hansen, Beth Daley, Jan E. Droege and Gina Jones for listening to my egotistical-sounding, enthusiastic monologue without any discouragement. I would love to recognize participants on the discussion forum on www.iSixSigma.com for their critically brutal feedback that made me work harder to clarify my framework. Cynicism is good!

During the validation phase, I would like to thank organizations that gave me the opportunity to present my framework in seminars or conferences. These organizations include the IPC, ASQ Northeastern Illinois Section (1212), American Productivity and Inventory Control Society (APICS), Scanlon Leadership Network, i-Solutions, and the Council for Competitiveness. I really appreciate Tony Hilvers, Marlyn Hyde, Tim Wilson, Majel Maes, Steve DuBrow, and David Attis, respectively, from these organizations.

While writing the book, I had tremendous help from Lisle Library. Without this library's collection of books on innovation and creativity, I could not even have started writing a book on innovation. While writing the chapter, *Brain Hardware and Mental Processes*, I had the toughest time for the longest duration. Special thanks to my friends' children Divya Jain, Kriti Goel, Surbhi Garg, and Krishna Gupta (my son) for loaning their school or college books to me. I also thank Dr. Eric Chudler for his permission to use figures from his website. Thanks to Ben Best for his work on his website, and Jeff Hawkins for his book *On Intelligence* for uncovering details of the brain. Without their help or work, I would be clueless about the brain's inner workings.

Contributors played a critical role in giving shape to this book by writing appropriate chapters. Given their demanding schedule, and professional and personal commitments, they each were able to contribute a chapter to the book. I am thankful to my contributor friends Jim Harrington, Hans Hansen, Abhai Johri, Alexis P. Goncalves, Rajeev Jain, Laurie LaMantia, Wayne Rothschild, and Justin Swindells.

Putting work together appears to be easy after writing all the chapters. This is an illusion. The work to put the book together begins after writing the chapters. The book would not have been completed without tremendous effort throughout the project from my associates and friends, including Shan Shanmugham for helping with Chapter 3, Shellie Tate for superb editing, Preeti Gupta for her ideas enhancing the cover design, Dan Pongetti for the beautiful book cover, and Arvin Sri for the rest of it. Working with these friends for so many years is a privilege.

I am thankful to Bob Anderson and Bob Carlson of the Illinois Institute of Technology for giving me the opportunity to share my work with them, and even more for letting me teach a course at IIT Chicago during the fall of 2006. I always learn from my students, whether at seminars or universities.

My special thanks to Randall T. Kempner, Vice President, and Deborah Wince-Smith, President of the Council of Competitiveness, for hearing me out in Washington D.C. Your feedback has been a tremendous source of strength for me to complete this book and was much needed during this journey full of support and skepticism.

I am blessed to know so many great people in my life. I am honored to still be in touch with my business hero, Bob Galvin, former chairman and CEO of Motorola. It is because of his goodness that I am able to present his thoughts not once but twice in my books, here and in the *Six Sigma Business Scorecard*. His contributions are beyond my thanks and remain a precious experience to me personally.

Over the years, I have learned that a person can brag about his accomplishment, but nothing can be accomplished without spiritual blessings. I am indebted to Brother Leo V. Ryan, former Dean who innovatively positioned the Kellstadt Graduate School of Business, for sharing his experience about the Leo V. Ryan Entrepreneurship Center. Personally, I am blessed to listen to the teachings of Swami Sharnananda Ji for inspiring his pupils to use more brain than material, which then inspired me to make my contribution to society.

I must accept that I am no student of Einstein, Newton, Galileo, Edison or Ford, because I am unable to fully comprehend their great work. However, I am a student of their thinking processes, which have led to great discoveries over the last five hundred years. My unplanned visit to the Boston Museum of Science to explore the Einstein Exhibits helped a great deal in creating the framework of innovation.

Finally, this book has been elevated by the Foreword from Dipak Jain, Dean of the Kellogg School of Management. I am grateful to him for reviewing my chapters on his vacation and for providing support and trust. His review means a lot to me, the contributors, and the book.

I have attempted to recognize individuals whose contributions have advanced my learning and the book. If anyone is ignored, please let me know at your earliest convenience, so I can correct the error in the next edition.

FOREWORD

From the moment a business begins, innovation is an integral part of its success. Competitive pressures mandate that organizations continually innovate to sustain profitable growth, but challenges exist in trying to maintain this momentum. In the knowledge age, various fields are increasingly interrelated, leading to a convergence of information in all areas.

In the past, innovation has been considered an art—dependent upon the people who make it happen and a relatively rare, unpredictable occurrence. As a result, too much time is taken in the development of new products and services. In the 21st century, we must change this process so that we become continual thinkers, capable of innovating on demand for mass customization. This requires that we understand innovation better and standardize the process for predictable results.

History shows that innovation is evolutionary and is a response to an unsolved problem or unexploited opportunity, which makes Praveen Gupta's *Business Innovation in the 21st Century* a valuable addition to the literature. His 25 years of business problem solving have given him the tools to develop a plausible framework that directs us to look at innovation in a different context. He is on the mark when he states a "networked individual" is a building block of innovation in the open business environment.

Several books have been written addressing various aspects of innovation, but common understanding has been limited to the level of "brainstorming" and creative tools. This is one of the first books that address various aspects of innovation from concepts to commercialization, guiding readers through the practiced experiences of various contributors to the book.

One of the challenges in institutionalizing innovation is to engage everyone in the organization intellectually. The proposed framework provides a theory, methodology, and measurement of success that is necessary if an organization is to accelerate the process.

I believe creativity happens, but I also believe that Mr. Gupta has helped uncover a way to make innovation happen.

Dipak Jain
Dean, Kellogg School of Management
April 15, 2006

INTRODUCTION

The purpose of this book is to provide comprehensive coverage of innovation-related processes and a new framework of innovation—a framework which is suitable for the Internet generation and the knowledge age. The goal of developing such a framework is to facilitate standardization of the innovation process so that results can be more predictable and so that innovation can be produced on demand.

The expectation of mass customization may require innovation on demand in real time. This expectation is a future "game" that business is going to have to play. Business, therefore, must participate in this knowledge innovation in order to make an impact. When a customer wants innovation on demand, a business must be prepared to meet that need. Innovation depends on being creative on the spot. Innovation is applied creativity.

Because humans are fundamentally creative beings who rarely do something the same way twice, teaching creativity to people is like singing to the choir. Creativity applied becomes an innovation. Being the youngest of twelve siblings, I used to be called "the procrastinator" because I would not follow any instructions and tried to assert myself by doing something different. Well, I recognized I could do something purposefully different when my "boss" wanted to develop an Input Protection test on semiconductor memory chips. I saw potential in it. I made some money as Motorola paid for publishing articles in magazines as an incentive for promoting creativity. Since then, "doing it differently" has become my mantra at work, and I have been trying different things.

Doing it differently means that I must be continually searching for ideas, i.e., looking for opportunities and information at the same time. I am like an information scavenger. I read junk mail more than really high-tech material for ideas. This book is an outcome of my learning, doing things differently, and solving problems creatively.

This journey led to my curiosity about how the brain functions so I could become an even more productive innovator. I have studied great innovators and scientists such as Newton, Einstein, Edison, Steve Jobs, Thomas Watson, Henry Ford, Tom Peters, Steven Covey, and Alfred Sloan. As is the case with every person mentioned above, in order to be successful, being innovative is required in order to differentiate oneself from the rest of the pack. In addition to studying great innovators, continually solving process problems led me to study business performance problems, which still is a challenge, because business is a complex process.

As I was completing my *The Six Sigma Performance Handbook*, I concluded that three 'thinkings' are required to really do good work. They are Process Thinking, Statistical Thinking, and Innovative Thinking. Further work in the area of innovation (60 plus books) and searches on the Internet led me to believe that a formal process for innovation did not exist.

Interestingly, the last fifteen years have been dominated by information and are leading to the knowledge age. Seeing the future through a maze of information, knowledge, innovation, improvement, customer demands, outsourcing, and globalization, I can see that we must become knowledge players.

Peter Drucker has been visionary in identifying and developing management principles. In 1985, he identified "new knowledge" as one of several sources of innovation, and knowledge innovation as a difficult type of innovation. Robert Weisberg has published extensive work on creativity and myths associated with it in his book, *Genius and Other Myths*. He says, ". . . given the large increases in our knowledge over the last decade or so, and the general consensus that seems to have developed among cognitive psychologists, a theory of creative thinking may not be too far away." The urge to learn more about the methods of innovation has existed for quite some time.

This book creates a framework to answer some of the questions about innovation. In order to institutionalize innovation, the contributors address various aspects of innovation from its history to its strategy to its implementation. Once the process of innovation is understood, and people realize their ability and the feasibility of innovation, one can see the near future, do wonders, and accelerate further innovation. However, a multi-disciplined approach utilizing experts is required to develop such a process.

This book has benefited from the expertise of many contributors who bring their specialized knowledge of different aspects of innovation to the reader. Accordingly, this book can be divided in three parts. "Part I: Evolving Innovation" looks into the historical aspects of evolutionary innovation, current or conventional tools and techniques, and future needs. The purpose of Part I is to bring the knowledge of innovation as an art up to date. "Part II: Understanding Innovation" presents the new understanding of innovation and the innovation process in the knowledge age. Part II expands the knowledge of innovation and introduces it as a science. "Part III: Institutionalizing Innovation" focuses on the implementation of various aspects of innovation in order to generate value. The purpose of Part III is to learn methods of adapting innovation in various organizations.

The background of the contributors is quite varied and makes the discussion in the book more diverse. The contributors come from a variety of backgrounds including human capital, IT consulting, law, banking leadership, business executive leadership, and an innovator with over 50 patents. Their rich experience is reflected in their chapters. To top it all, an interview with Bob Galvin, former Chairman and CEO of Motorola, regarding his innovation practices is included in the book. Under Mr. Galvin's leadership, Motorola implemented sound innovation processes and realized significant growth.

Many recently-published books use case studies in order to make their point. Utilizing case studies in a book has its advantages and disadvantages. The main advantage is that the reader can quickly see how an idea or process worked at a company and then either forget it or try to emulate it. Emulating it is probably not the ideal approach for maintaining personal creativity. Emulating implementation of a methodology in one organization most likely will lead to similar pitfalls the "case" company faced and prevent new innovative thinking for effective results. I have been told many "case" companies in renowned books or journals have failed or have never been favorably highlighted again.

Incorporating cases into a study of innovation is very anti-innovative, because it stifles the natural creativity of the reader and practitioner. We must first learn to understand innovation concepts and apply them creatively in our organization. If one understands the intent of the new framework, theory, and methodology, the implementation phase becomes a challenging yet rewarding task. If we do not understand the concepts of innovation, we just emulate the innovation process of a 'case' company. This approach is unlikely to work, however, given the "people potential" of a different organization.

Readers can review Part I if they want to be tuned into the concepts of innovation, read Part II to learn the science of innovation, and study Part III to institutionalize innovation. Eventually, once a person reads the book and understands the framework for innovation, the reader should just let it sink in first before reacting or getting busy. Innovation occurs in a brain working in a networked environment, so we must allow time for the brain to absorb the new knowledge, build some anchors, and develop speedy innovation practices. Everyone can be innovatively innovative. Go for it!

Reading Roadmap

Intent	Strategic	Operational	Special
Audience	Executives	General	Technical
Chapters			
History of Innovation		X	
Creativity and Innovation		X	
The Conventional Tools of Creativity		X	
Innovation in the Information Age	X		
Need for Innovation on Demand		X	
Brain Hardware and Innovation			X
Framework for Innovation	X	X	
Room for Innovation		X	
Innovation Deployment		X	
Measures of Innovation	X	X	
Innovation in Service		X	
Protecting the Innovation		X	
Commercializing Innovation		X	
Managing Innovation	X		
Final Thoughts. Wisdom of Innovation	X		

PART I

Evolving Innovation

HISTORY OF INNOVATION

Praveen Gupta

In the chronology of social development, when human beings started to acquire assets, competitive desires evolved—a desire to have more, a desire to have something better, a desire to have things that make life easier. When these desires became strong enough, they transformed into a need or necessity, and as the old cliché goes, "necessity is the mother of all inventions." Historically, knowledge was limited to a few. However, in today's knowledge age, where information is shared widely, future innovation will result from individual and collaborative discoveries at an increasingly faster rate. Studies show that innovation is built on past knowledge and continuous experimentation. Rather than accepting it as an ad-hoc process with unknown outcomes, innovation can be developed into a structured process and a more predictable system.

Innovation has always been a part of mankind. Since the discovery of fire by rubbing two stones together, humans have been innovating. Innovation is probably the oldest known process; in other words, innovation is an extension of a person's creativity. We have always used our innate skills to create many new things and to help mankind.

Imagine when the human evolved and discovered fire. What was the knowledge level then based on what we know today? What was the level of excitement at the discovery of fire? As people gain new understanding by trial and error, they transform it into new knowledge and then use that knowledge to gain new understanding, discover more unknowns, and

1

become even more curious. Thus the cycle of experimentation, knowledge, and innovation continually repeats. The outcome of the knowledge-experience cycle has led to continual creativity and innovation.

HISTORY AND EVOLUTION OF KNOWLEDGE: FROM STONE DAGGERS TO METAL

Before discovering fire, humans discovered simple rocks that could be used as tools. Getting ideas from human or animal teeth, a thought of a dagger could have arisen, and so daggers of stone were made. A dagger provided protection from animals and probably was used as a tool to prepare cold food which was then warmed by the sun's heat. Daggers could easily have evolved into knives and spears. These tools could be used to tame animals or even for hand-to-hand fighting.

Humans discovered fire more than 50,000 years ago. Fire, which could be very destructive if not controlled, could be a great friend when controlled. The discovery of fire led to further human knowledge, as the fire could be used for making tools, keeping humans warm, keeping animals away, cooking meals, lighting dark caves, or even melting ice. Therefore, the discovery of fire could be considered a great breakthrough in human evolution because it was critical for survival.

How could the early humans or hominids get an idea about fire? They must have observed fire caused by lightning, or sun heat, or volcanic eruptions. They could have even observed fire while throwing rocks which produced sparks when they hit other rocks. The discovery of fire led to humans thinking about how to use fire and how to protect themselves from it.

Thousands of years later, humans did invent the bow. The idea of a bow could have come from tree branches loaded with fruit. In thunderstorms or high winds, tree branches often throw their fruit far away. The tree branches may have been the catalyst for the invention of slings for throwing rocks, and slings led to bows to launch arrow-like spears. The arrow could be considered an evolution of spears adapted to work with bows for throwing longer distances.

The discovery of daggers, knives, fire and bows and arrows may have led to the preparation of warm meals. Warm meals resulted in warmer bodies and may have led to the need for clothes to satisfy the demand for warmth. Clothes made out of grass and roots evolved to clothes made of animal skins with the help of a needle. Therefore, the discovery of the needle was a breakthrough. The early needles were like a hook to stitch two pieces of skin or fabric to replace the series of knots previously used to put two pieces together. The knots could have been discovered from natural entanglements of long string-like objects, or even tree branches or bushes.

Early civilization appears to be based on the seven metals, as the remaining known metals were discovered since the 13th century. The seven metals are gold, copper, silver, lead, tin, iron and mercury. Early tools and weapons were made of copper, which was discovered around 4000 BC, and tin and iron were discovered around 1500 BC. The discovery of copper was more significant, as the first set of tools, implements and weapons were made of copper. Early applications of copper were made with hammer and chisel.

Copper smelting was probably learned while throwing copper waste into fire. The first copper-smelted artifacts were found in the form of rings, bracelets, chisels, and weapons about 500 years after the discovery of copper. By the 17th century, an additional five metals were isolated, which are platinum, antimony, bismuth, zinc, and arsenic. By this time, metallurgy was a well-developed discipline. Post 17th century discovery of metals accelerated as twelve new metals were discovered in the 18th century.

HISTORY AND EVOLUTION OF KNOWLEDGE: LANGUAGE, NUMERALS AND ART

Panini (6th century BC), an Indian mathematician, developed a theory of phonetics, phonology, and morphology, and provided formal production rules and definitions describing Sanskrit grammar in his treatise called *Asthadhyayi*. Basic elements such as vowels and consonants, and parts of speech such as nouns and verbs, were placed in classes. The construction of compound words and sentences was elaborated through ordered rules operating on underlying structures in a manner similar to formal language theory.

In the modern world, around the 12th century, Raymundus Lullus invented the logical machine, *Ars Combinatoria*, in a deep crisis of communication. Lullus started a revolution of formalistic thinking to produce declarations in a mechanical manner. He founded the concept of organizational thinking by constructing a paper-machine to combine language and geometrical figures (represented by signs and letters) for capturing various declarations the human mind could conceive. It consisted of three circular paper disks that were fixed on an axis on which they could be turned for producing possible combinations of letters and symbols, thus leading to the development of deciphering signs, creating organized thoughts and the processes of decoding and encoding. In other words, language was born out of graphic representations of signs and studying their associated patterns.

Language evolved based on natural sounds and representations as well as circumstantial human body expressions (i.e., pain, anger or joy). Physical gestures led to oral expressions. Humans have evolved upright teeth, small mouths to make sounds, intricate muscles in lips, a very flexible tongue, and a resonating larynx. The desire to express personal feelings led to communication or the interactive language, and describing natural phenomena led to the development of transactional language or communicating observations, knowledge, or skills. This transactional language led to the development of a written form of language. In other words, written language must have evolved from the spoken word, pictograms, syllabic writing, and alphabetic writing.

In India, the decimal system existed in the pre-1000 BC era and migrated to the Middle East through the translation of Indian literature. Even though the Indian-Arab numerals had been in existence in 300 BC, the use of numerals began to grow in Spain in 900 AD. Leonardo Fibonacci introduced the Indian-Arabic numerals to Europe in 1200 AD. However, it was not until the 15th century that the European tradesmen, bookkeepers, and surveyors started using Arabic numbers instead of Roman numerals.

With the understanding of natural phenomena in mathematical terms as well as through language, society moved into innovating through art first by building large temples, churches, palaces, pillars, tombs, pyramids, and monuments. These monuments demonstrated innovative thinking and its manifestation through combining various patterns and through the evolution and discovery of new structures.

As society grew, population increased and grouping took place. With the accumulation of assets, competitive desires evolved and the race to discover more progressed. This interaction probably led to comparison and competition to have more than the other. All this evolution eventually leads to greed, wars, destruction, renewal, and a continual, growing demand for innovation.

EVOLUTIONARY TIMELINE

Figure 1.1, Understanding Art and Science, shows that looking at objects, rituals or practices creates curiosity, a questioning mind, and a decision to like or dislike the input. If an activity is desirable, attempts to replicate it are probable. If similar desirable results are achieved, the activity is reproduced for more and better results. When an activity is repeated without recognizing its details, it becomes an art; thus it possesses more perceptual and tacit knowledge.

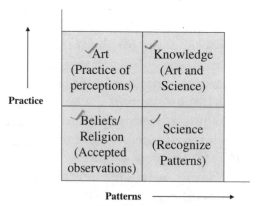

FIGURE 1.1. Understanding Art and Science

A belief is formed and is accepted when its understanding and practice are not questioned. A set of beliefs may define a religion for most of us. When an activity is understood through its patterns and repeated practice, it represents a methodical study of an observation. Science is defined as a methodical study of a subject that leads to an understanding of cause and effect. When the art and science of a subject are understood, greater competence is achieved and a certain level of knowledge is acquired. Therefore, a person is knowledgeable when he or she understands the art and science of a topic.

Figure 1.2, Timeline of Innovation, displays human evolution through growth in art, science, and language. History shows that art and science go together and sometimes build on each other. Since the discovery of fire and through general observation, reflection, and the desire to do something different, humans have innovated from bows and arrows to rockets and the space shuttle. We have progressed from simply being able to stand straight to flying in space, from throwing a rock to firing rockets in space, and from launching spears to launching nuclear warheads.

With this evolution in knowledge, expertise has evolved and diversified to that possessed by philosophers, priests, artists, mathematicians, scientists, engineers and the SMEs (Subject Matter Experts). One can see that discoveries are always built on prior but insufficient knowledge. An observation counter to existing knowledge generates more questions and answers to those questions create more knowledge, and results in more discoveries.

The Information Age, which started in the 1980s, lasted to the end of the 20th century. The impetus for the information age began with an observation by Thomas Edison, which led to development of the diode, transistor, semiconductor chip, computers, and super computers. The information generated during the last two decades is leading to more questions about utilizing that information, providing business intelligence, and raising more questions about potential new discoveries.

In the 21st century, people are at a much more advanced stage of materials, information, science, art, and language. The new era of information mining is requiring new business intel-

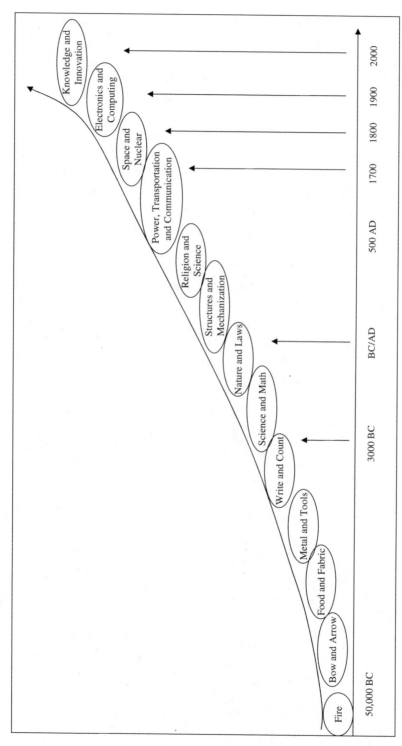

FIGURE 1.2. Timeline of Innovation (Not Plotted to Scale)

ligence about optimizing business performance. The process of quickly creating new knowledge from existing knowledge demands a better understanding of the innovation process. Therefore the knowledge age, which has already arrived, must accelerate innovation through the collaboration and intellectual involvement of humans.

Such innovation will accelerate knowledge creation as well as product and service development. The efficiency of innovation will require clustering, enhanced team-intelligence, intellectual engagement, and better understanding of the innovation process. Ultimately, humans are the building block of innovation, whether the innovation is occurring due to collaboration involving individual innovators, or is distributed to multinational innovation teams. In other words, future innovation results from individual and collaborative discoveries at an increasingly faster rate.

Figure 1.3, Influencing Innovations of Recent Centuries, illustrates that by the end of the 18th century, a turning point occurred that led to a higher number of discoveries or innovations. The 18th century was a renaissance of innovation when people started to think about a system of innovation. The establishment of the U.S. Patent and Trademark system evolved from the understanding that innovation would be a critical component of the American economy. The founding fathers must have been

18th Century	19th Century	20th Century
Sewing Machine	Refrigerator	Air Conditioner
Electronic Battery	**Telephone/Fax Machine**	**Theory of Relativity**
U.S. Patent System	Electromagnetic Induction	**Mass Production System**
	Elevator	**Traffic Light**
	Dynamite	Television
	Typewriter	Xerography
	Wireless telegraph	Nuclear Reaction
	Light bulb, Electric Power	Transistor/Integrated Circuits
	Business management principles	Digital computers/PCs
	Motorcycle/Automobile/Airplane	**Internet/World Wide Web**

FIGURE 1.3. Influencing Innovations of Recent Centuries

thinking that innovation would become a normal phenomenon, and they prepared to handle it, protect it, and accelerate it.

The fact that the 19th century led to more innovations than the 18th century is evidence of a mature understanding of the innovation process. During the 19th century, many products addressing improvement in human existential capabilities were developed such as the refrigerator, transportation, and telecommunications. Significant innovation also occurred in the arenas of thermodynamics, electromagnetics, gravity and electricity, which became the seedbeds of Einstein's discoveries.

The 20th century led to the commercialization of many scientific discoveries of the prior century. Major commercialization was realized in the fields of nuclear physics, mass production systems, telecommunications, xerography and photography, television, and outer space expeditions. The most significant discoveries of the 20th century were achieved by Einstein and Edison. Einstein developed the theory of relativity, and Edison discovered Edison's effect. Einstein's theory helped people understand the basic dimensions of the universe, time and space. Edison's effect led to the evolution of the electronics industry, which was started with the development of diodes. The diodes led to triodes, pentodes, transistors, semiconductor chips, integrated circuits, and the computer. Computers which helped increase the pace of information processing and analysis became the catalyst for the advent of the information age.

THE WORLD OF INNOVATION

Understanding the roles of, and methods deployed by, Einstein and Edison requires looking at their accomplishments in a "big-picture" context. Figure 1.4, Fundamental Innovations, and Figure 1.5, Business Innovations, enable us to review scientific evolutions at a higher level from the beginning. These two figures display the role of each great innovator in the context of the big picture, and they also provide a comparative analysis.

The work of Galileo, Newton, Edison, Ford, and Einstein represents a period from the 16th century to the 20th century,

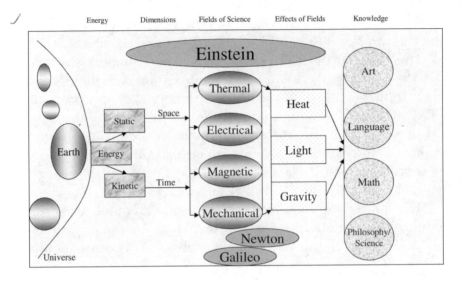

FIGURE 1.4. Fundamental Innovations

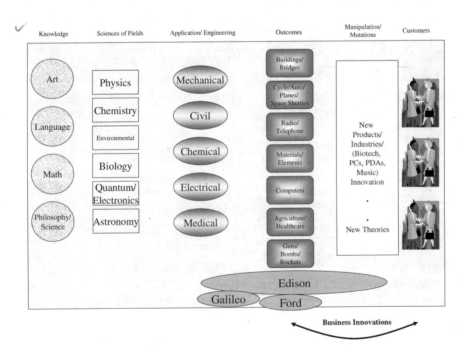

FIGURE 1.5. Business Innovations

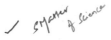

which was an era of super-scientific discoveries. These innovators either made significant contributions or recognized major natural phenomena that helped humans understand the universe better, thus leading to extensive further discoveries.

Figure 1.4, Fundamental Innovations, begins with the so-called Big Bang theory, which represents time t = 0. In other words, at some point in time the universe was formed. It grew tremendously in its first few fractions of a second of existence. The universe is associated with some level of energy, which is split into kinetic and static energy. Kinetic energy relates to the motion of objects in planets. Kinetic energy is associated with the dimension of time, which represents the change in distance with respect to time. Static energy relates to the space or distance between objects, including that of planets.

The four basic fields of nature are mechanical, electrical, magnetic, and thermal. The first three fields happen to be associated with the intrinsic properties of objects. The mechanical field represents static (location) and dynamic (movement) positions of objects, while the electrical and magnetic fields represent stored forms of energy due to the structure of objects. The thermal field represents a conversion of energy from one form to the other due to changes in space, position, size of the objects, and inherent (unknown) properties of the material.

These fields of science manifest themselves in the form of heat, light and gravity. Gravity is a force between two objects dependent on the total energy associated with larger objects, or objects with higher specific density. The larger object cancels out the gravitational effect of the small object and applies net gravitational effect on the smaller object. Light and heat represent a spectrum of energies when the state of the material is changed. For example, the Sun is emitting both heat and light based on the conversion of energy due to a change in the state of material with which the Sun is composed. The ratio of heat and light depends upon the nature of objects and their extent of transformation from one form to the other.

An assumption being made here is that the universe preceded living organisms. Humans evolved from other species as the universe and living organisms co-existed over time. People evolved over time and started observing the various effects of these fields. They could feel the heat, see the light, and walk a distance. People tried to understand and communicate these observations about nature's effects. When they could not explain them, these unexplained observations became beliefs, and their view of the world became their philosophy.

As they learned more about nature, they started to refine their model of the world by observing patterns. They also developed different models of communication. For example, when people quantify an effect, it becomes math. When they describe observations of objects, they utilize language, and when they display what they see, we call it art. In other words, art, language, and math are all human expressions of what is going on in the world or universe. Knowledge is a combination of art, language and math. The process of exploring objects utilizing art, language, and math is called science, and science is a collection of knowledge about various objects.

Figures 1.4 and 1.5 show the types of innovation various innovators produced. For example, Newton studied in the field of mechanics, while Galileo worked at the science level as well as the product level. Galileo developed products such as the pump and compass, while Ford was more of an innovator in business processes such as lean manufacturing. Einstein focused more on the basic science of fields. Edison was committed to developing new products and processes. Edison have filed and received the most patents (over one thousand) in the world.

Figure 1.5, Business Innovations, represents an iterative utilization of known effects and knowledge of the universe for creating new knowledge. From knowledge of the universe in terms of art, language, and math (collectively called science), humans created studies of major aspects of the universe for further exploration. These studies promote ongoing learning regarding the universe and lead to applications for the betterment of mankind. In other words, the methods of applying science to develop new products define various engineering branches such as mechanical, electrical, materials, civil and chemical.

Engineers develop new products that are used by people to provide new solutions for making life better (i.e., creating new knowledge). People adapt to these conveniences and demand more products or services. The cycle of demand and supply begins and thus business starts. Businesses package these solutions for customers based on the current knowledge and supply them to customers in return for considerations in kind or currency.

- The development of new processes, products, services or solutions to continually create more value for customers is called business innovation. Business innovation occurs when a process or product offers such a significant change in creating value that customers benefit from the increased value. Business innovation can disrupt the current method of utilizing a solution or doing business; business innovation can even create new business or possibly a new industry.

As Figure 1.5 shows, business innovation occurs through the mutation of various dimensions of a solution (in the scientific realm) which promotes better-performing solutions or applications for us in our everyday lives. New products, new services, and new solutions are continually developed as a result of these mutations. Some mutations lead to an incremental change in performance, while others cause a radical change in performance. The extent of change in performance of the new product or service determines the type of innovation (i.e., incremental or radical).

These new products or solutions enable the creation of newer knowledge and add to the existing knowledge base. Most product innovations occur between the customer and the engineering or development teams. Sometimes, however, new cumulative knowledge leads to the discovery of new engineering solutions, or even a new science. Figure 1.6, Paths to Continual Innovation, shows that innovation can occur at all levels. The farther the path of innovation goes, however, the harder it becomes to innovate, and the lapsed time between innovations is longer.

While Einstein published four major papers in 1905, and Edison planned a patent per week, now more business innovations are to be expected compared with basic scientific discoveries. Many business innovations occur on a daily basis. As mentioned previously, every once in a while, however, a new field of knowledge, such as fractals, is created. In addition, new engineering disciplines such as software engineering, bio-tech

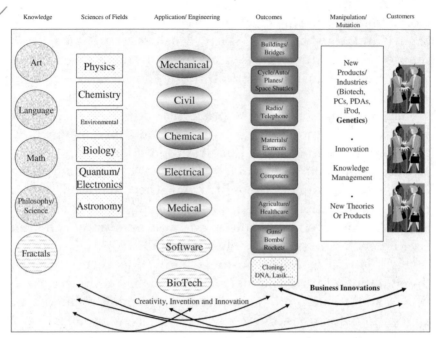

FIGURE 1.6. Paths to Continual Innovation

engineering and human engineering arrive on the scene. These new disciplines lead to new products such as Lasik surgery, DNA testing and cloning.

GREAT INNOVATORS: FROM GALILEO TO EDISON

To establish a standard process of innovation, observation of the details of innovations from great innovators should occur. Figures 1.7a, and 1.7b Great Innovators, depict a life-time engagement in discovery and innovation from some of the greatest individuals who took a unique approach to their accomplishments. Galileo, Newton, Einstein, Ford, and Edison all excelled in scientific, technical, industrial and business innovation.

Galileo was an Italian astronomer and mathematician. He studied the works of Euclid and Archimedes, developed a pump for raising water and a high quality refractory telescope to study

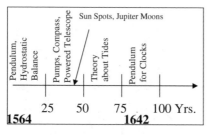

Galileo's Innovative Path

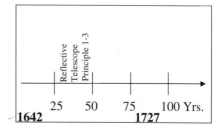

Newton's Innovative Path

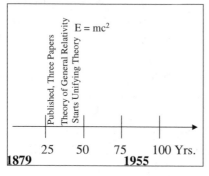

Einstein's Innovative Path

FIGURE 1.7a. Great Innovators

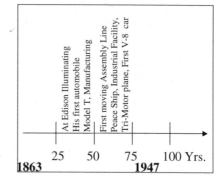

Ford's Innovative Path

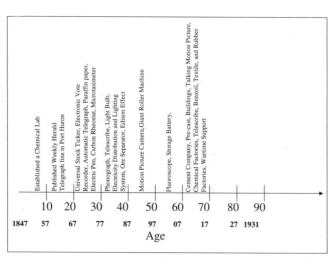

Edison's Innovative Path

FIGURE 1.7b. Edison - The Greatest Innovator

stars, invented a geometric compass and a thermoscope (i.e., thermometer), developed hydrostatic balances, designed pendulums, and applied the concept of pendulums to clocks. He observed that objects of different densities achieved the same rate over an inclined plane. Galileo utilized his own telescope to make celestial observations about sunspots and star formations, and he discovered new stars and their movement.

Galileo was a mathematician as well as a craftsman. He made great observations and posited bold theories when the war between religion and science was at its peak. He was convicted by religious leadership and imprisoned in his house for his belief that Earth rotates around the sun. Risk, part of an innovator's life, was experienced first-hand by Galileo. Galileo explored and innovated for about 55 years in a variety of fields, even during his imprisonment.

Galileo died on January 8, 1642, and Isaac Newton was born on December 25 of the same year. Newton studied mathematics and started to make his mathematical entries in 1664, completing his legendry publication *Anni Mirabilis* in 1666. In 1669, Newton developed a reflecting telescope and studied light and colors. Newton then studied planetary motion and observed Halley's Comet in 1682.

Newton's best published work is *The Mathematical Principles of Natural Philosophy*, or *The Principia*. Newton proved Kepler's Third Law about the motion of elliptical bodies in orbits. *The Principia*, completed in 1686, became the most comprehensive book on forces of gravity and motion between objects. *The Principia* includes definitions, rules, laws of momentum (mass and motion), forces (inertial, impressed and centripetal), and definitions of time, space and motion.

Newton then applied his laws of motion to the motion of planets, moons, and comets as well as to the behavior of Earth's tides. Newton built on the work of Kepler, Galileo and others. *Principia* was the most scientific approach to studying physics at that time. In Newton's own words, he researched his topics of interest and advanced the prior work. He also recognized his own possible "defects" in such a difficult subject and encouraged further investigation by his readers. Newton was a trained scientist who studied natural phenomena in a systemic fashion and who knew how to innovate methodically.

At the age of five years, Einstein was mesmerized by the movement of the needle of his personal magnetic compass. The continual orientation of the needle northward pointed to the existence of some invisible forces somewhere. At the age of 12, when he studied Euclidean plane geometry, he concluded that certain truths can be proven without doubt and with a sense of certainty. Einstein developed an uncanny ability to concentrate on topics leading to fundamentals and to clear a multitude of distractions out of his mind.

Einstein generally was a good student, was outstanding in mathematics, and hated memorization. He enjoyed studying mathematics, physics, and philosophy. He was even considered a distraction in his class at times.

In 1901, Einstein had a temporary teaching assignment and worked in the patent office in Bern from 1902 to 1909. While in the patent office, he wrote on theoretical physics on his own without being associated with a science community. During this time, he also earned his Ph.D. from the University of Zurich in 1905, which happened to be the year he made history. He revealed his breakthrough research in March, May, June and September of 1905. In March, he published his theory of light quanta (i.e., the particles of energy versus the conventionally accepted theory of light as oscillating electromagnetic waves). In May 1905, Einstein submitted his theory of kinetic energy explaining the so-called Brownian motion, which reinforced the kinetic theory and helped in the study of atom movement. Actually, Einstein's light quanta theory was based on his experiments on particles.

In June 1905, Einstein published his work unifying the application of the relativity principle between electromagnetic waves and motion. Earlier Galileo and Newton had studied relativity for mechanical objects, while Maxwell and Lorentz studied the effects of relativity on electromagnetic effects. Their electromagnetic theory predicted that the velocity of light would show the effects of motion, but they could not prove it in the lab.

Einstein theorized that both mechanical and electromagnetic effects would be affected by the principle of relativity. In September 1905, continuing his work on the principle of relativity, Einstein reported that if a body emits certain energy, the mass of the body must be reduced proportionately. The relationship between mass and energy was defined by the famous equation, $E = mc^2$. Einstein unified interactions among particle motion, optics, and electromagnetic waves based on the prior work of Galileo, Newton, Maxwell, Lorentz, and many more.

Henry Ford, born in 1863, had an interest in mechanical activities. He worked with steam engines, farm equipment, and factory equipment. He had a sawmill business in early adulthood. Later he joined the Edison Illuminating Company where he became chief engineer in two years. During that time he experimented in internal combustion engines, which led to the development of his own self-propelled quadricycle and the formation of the Ford Motor Company in 1903.

Ford became a social entrepreneur and improved his manufacturing processes to produce cars at a reasonable price. He combined precision machining, standardized processes, interchangeable parts, division of labor, and assembly-line manufacturing, where the product passes by the worker for assembly. On the assembly line, Ford used conveyor belts, conducted time studies, and accelerated the Industrial Revolution. He significantly reduced the cost of manufacturing for his famous Model T cars.

Around 1920, Ford built the world's largest industrial complex which included a steel mill, glass factory, automobile assembly line, rolling mills, forges, assembly shops and foundries. Basically all processes from refining raw material to the finished automobile were now performed at one plant. Ford created the concept of mass production and all the components associated with it.

As a social entrepreneur, Ford cared about his employees' lives at home and at work, built cottage industries in rural areas, established schools in several areas country-wide, and created the Henry Ford museum to preserve past innovations for future generations. At the Ford museum, the evolutionary nature of the innovation process is seen in living color when viewing old methods of washing clothes compared to auto-mated washing/drying using the latest washers and dryers. Clearly innovations are built on what came before or on obser-vations of an existing natural fact.

Thomas Alva Edison was born in 1847, a few years ahead of Ford. Edison had only about three months of formal education and became the greatest innovator of all time. He established a chemical laboratory at the age of 10. Edison's exemplary discov-ery was the light bulb, and hence the entire lighting industry was born. Edison's greatest contribution was the first practical electric lighting.

Edison invented the phonograph, telegraph and telephony components, such as the carbon microphone, motion picture camera, electronic vote recorder, and the universal stock ticker. Edison also assisted on the production of the typewriter, elec-tric pen, paraffin paper for wrapping candies, wireless telegra-phy, dictating machines, shaving machines, improved electric railways, roller machines to break large masses of rocks, fluor-oscope, storage battery, Portland cement, electric motor, pho-nograph, kinetophone (sound and motion), and carbolic acid for explosives in World War I.

Edison figured out the innovation process in his early child-hood. Edison epitomized the innovation process by combining scientific, industrial, and business innovation through his de-sire for continual innovation and growth. Edison continually grew professionally, personally, and financially through his endeavors. He was a gifted innovator who believed in working hard, learning from everything he did, and improving and in-novating on everything he did. As a result, he expanded experi-ence more, learned more, and innovated continually. He was so fascinated by creating new products that he set up the first modern-age research laboratory where he facilitated and accel-erated innovation.

Edison really mastered the innovation process. He had over 1,000 patents—the most issued to an individual. His famous quote: "Genius is one percent inspiration and 99 percent per-spiration" still reverberates throughout the world. Edison loved physical and intellectual work. His heart, head, and body must have been busy all the time. He believed that innovating new things was a good task to undertake in order to gain fame and fortune while also benefiting society.

NATIONAL INNOVATION INITIATIVES

In response to the growing need for innovation, several countries have already developed national policy regarding innovation. These policies were developed based on status of innovation, expected economic growth, and necessary infrastructure to achieve desired results in terms of new industries, new products, or new services.

For example, The Organisation for Economic Cooperation and Development (OECD) is a group of 30 countries that share democratic governments and market economies. OECD's work addresses social and economic issues from macroeconomics to trade, education, development, science and innovation. OECD's Science, Technology, and Innovation initiative looks into developments in the area of innovation policies in member countries and provides direction based on surveys, research, and benchmarking.

The Lisbon European Council launched the European Innovation Scoreboard (EIS) in 2000 to support the European Union's (EU) goal of becoming the most competitive and knowledge-based economy in the world. The EU's industrial policy emphasizes innovation. The EU's Innovation Policy stresses entrepreneurial innovation or knowledge innovation. In 2003, the EU established a framework of common objectives for strengthening innovation. The EU's Innovation Scorecard and European Trendchart on Innovation provide analysis of national innovation policies and benchmarking. The EIS provides overall national innovation performance, while the Trendchart policy database provides country reports and continual lessons learned on specific issues.

Iran's government has formulated a project to study innovation. This project, called the National Innovation System, was created for visualizing effective policy measures, exploiting the potential of emerging sectors, reducing its dependence on the oil sector, and highlighting potential opportunities for growth.

China's new science and technology policy statement calls for greater efforts to instill a spirit of innovation in its society. Its policy also addresses the need to develop and commercialize high technology, such as software. China's drivers for innovation include eradication of poverty, improvement in the quality of life, and strengthening its defense. The Chinese are using innovation, science, and technology to fend off so-called pseudoscientific ideas or cults.

India recently established a National Knowledge Commission for three years, comprising of experts in various fields. These experts advise India's national leadership on policy matters primarily regarding its education system and how to build India's "knowledge power" in the world. Its focus, the Knowledge Pentagon, includes excellence in education, research in science and technology, intellectual property management, knowledge application in agriculture, and knowledge capabilities in making the government more effective and transparent. To ensure collaboration of various government organizations, India's Prime Minister heads the Steering Group overseeing the Commission's work.

In the U.S., several organizations are raising awareness of the need for innovation. To meet the slide in its economic performance, a nongovernmental organization, the Council of Competitiveness (representing industry, education and labor), was formed. Its guiding mission is to set an action agenda that drives economic growth and raises the standard of living for all Americans. The Council's Center for Regional Innovation facilitates innovation-driven economic growth through benchmarking.

UNDERSTANDING INNOVATION MODELS

After reviewing work on developing national policies and corporate strategies, and speaking with professors at leading universities, it appears that the main methods of innovation are collaboration, networking, and brainstorming. Some of the innovation classes show that the innovation process consists of interaction among people and sheds no light as to what the interaction does. One of the proposed theories, based on extensive research led by Prof. Michael Porter of Harvard University, calls for the creation of Clusters of Innovation. The U.S. Council of Competitiveness proposes to oversee such regional clusters of innovation. Accordingly, a cluster is defined as a geographical concentration of competing and cooperating companies, suppliers, service providers, and associated institutions. One of the examples cited as a well-recognized cluster of innovation is Silicon Valley.

The Clusters of Innovation appear to be an observation about how strategically resources are committed for regional economic development. They shed little or no light, however, on the process of innovation—in other words, how the clusters of innovation create innovative solutions, how efficient these clusters of innovations are, and how repeatable and reproducible these clusters of innovations can be.

Mere observations and statistical correlations between these clusters and evidence of innovations do not establish a causative relationship between the cluster and innovation. We must understand the building block of innovation so that it can then be institutionalized to accelerate innovation. Normally, these clusters have evolved over time, with significant resources, and were initiated by a larger institution. At a smaller level, these clusters could even be compared with incubators that promote and support entrepreneurship. The question still remains, however, regarding how to innovate efficiently and create new knowledge and new opportunities for continual growth and scientific evolution.

KNOWLEDGE INNOVATION

With the advent of the Internet, information is becoming available quickly. The Internet has already provided tools to collaborate among people globally; in other words, we can have clusters of innovation without clustering geographically. Moreover, the rate new information is being added on the Internet is itself exploding exponentially.

This information explosion looks like it will continue, and the protection of intellectual property could become trivial in many cases. The future need for innovation on demand in real time will mandate that corporations create new solutions to meet customer needs and then quickly move on to create other innovative new solutions to meet the next wave of customer needs. In other words, the laws to protect intellectual property will have to be re-examined as the rate of innovation increases. Current slow bureaucratic systems will not be able to keep up with the explosion of innovative solutions and related intellectual property.

In the upcoming and exploding knowledge age, customer-supplier relationships will appear to be very close, interdependent and insistent on innovative solutions. Current application of lean thinking demonstrates that business systems will be designed to produce to order, rather than produce to create demand. In such a scenario, if each item shipped by a company is unique, the innovation process must be institutionalized throughout corporations.

The expected extent of innovation goes far beyond development of products and services. Instead future customer demands will mandate innovation at every level in an organization in order to be able to serve customers and grow profitably. In other words, this type of innovation outlook is how a company can grow making millions of unique widgets.

Utterback, in *Dynamics of Innovation*, explores the relationship between process and product innovation. Product innovation leads to process innovations and vice versa. However, the success of a business is not ensured by just product or process innovation. Sometimes, very innovative products do not live up to their potential, and simple innovations exceed their expectations.

Figure 1.8, Dynamics of Innovations, shows paths to innovation beyond process and products. A process is composed of activities which are outcomes of ideas. The resulting products, when sold to customers, bring more business; that is what the corporations are all about. In other words, if there are no ideas, there will be no new activities or experiments, and if no

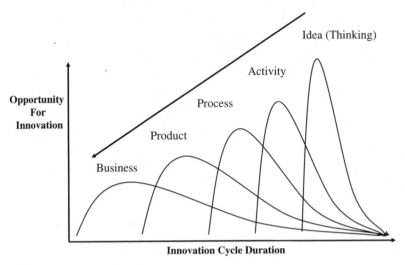

FIGURE 1.8. Dynamics of Innovations

new activities are performed, new processes will not evolve. Fewer new products or solutions, therefore, will be available for fueling business growth. Lots of ideas are needed on a continual basis to make the idea-to-business-cycle work. Dauphais, in *Straight from the CEO,* confirms this and likewise mentions a lot of ideas are needed to arrive at one new innovative product. Dauphais also asserts that businesses need many new products to hit a homerun.

If a business is internally driven to create demand, a continual stream of ideas leading to new products must be flowing. In other words, innovation begins with ideas. To generate a stream of ideas, a corporation must utilize all its intellectual resources (i.e., its entire workforce). When a corporation is engaging in *customer-driven* innovation on demand, speed of innovation counts. In some cases, the current sequential innovation process tends to be ineffective; instead the knowledge age innovation process is the place where many minds collaborate to create an effective solution on demand. In either case, many minds in a corporation must be involved to generate new ideas for innovative products and solutions. Excellence in idea management will become a corporate imperative in order to grow profitably.

INSTITUTIONALIZING INNOVATION

Involving all employees requires that we understand how to utilize their intellectual potential in creating the new intellectual property. The process must be standardized to a great extent, with some exceptions. Therefore, the innovation process must be well understood. One cannot accelerate innovation as an art; it must become science in order to accelerate. The paradigm of innovation must be that it is a science as much as it is an art. In order to innovate with a higher probability of success, corporations must look into various elements of the innovation process and practice them just like any other existing standard process.

In the early years after the discovery of electricity, businesses used to have a Chief Electricity Officer. As electricity matured and commoditized, it became a utility, and no longer was there a need for the electricity officer. Similarly in the information age, we have a Chief Information Officer to glean tons of information. To utilize the information, businesses now are trying to extract intelligence, so business intelligence is becoming an important issue that people are addressing through dashboards and scorecards. However, the application of business intelligence is to create new knowledge and new solutions. Therefore, the position of Chief Innovation Officer is going to become a natural evolution of the changing business model.

Studies show that innovation is built on the past. In other words, all innovative solutions are based on past knowledge, continual experimentation, and extension of this past knowledge and experimentation. The *process* of innovation appears to be evolutionary in nature as well. People must understand this evolutionary nature of the innovation process, open new doors to new insights in the world around them (or even the universe), and search for new solutions. Einstein implied that all innovations are merely discoveries. Therefore, people must continually strive to discover new aspects of business and the world. Once people accept that innovation is a result of the discovery process, not a subconscious effort, the process of innovation can be easily understood and established as a predictable system.

TAKE AWAY

1. All discoveries were the result of some observed phenomenon.
2. Discoveries are always built on prior knowledge.
3. Knowledge generates more questions, creates more knowledge, and produces more discoveries.
4. Learning from the work of great innovators, such as Galileo, Newton, Edison, Ford, and Einstein (representing the thinking of five centuries), is essential in order to become a great innovator.
5. Einstein mastered the thinking process, while Edison perfected the innovation process.
6. Many countries have established a national policy for promoting innovation to maintain economic progress in the 21st century.
7. At present the perceived understanding of the innovation process relates to brainstorming and collaboration.
8. The concept of 'clusters of innovation' is an observation of collaborative innovation in resource-rich areas. This concept, however, does not take into account the process of innovation.
9. To meet growing customers' demands for innovative solutions, the innovation process must be standardized.
10. Innovation begins with an idea; thus excellence in employee idea management will be the competitive edge and catalyst for continual innovation.
11. In order to institutionalize the innovation process, it must be understood better.
12. The innovation process is a process of discovery and is evolutionary in nature.

CREATIVITY AND INNOVATION

Hans Hansen

While examples of creative individuals abound, creativity is significantly enhanced through interaction and collaboration between people, ideas and the environment. Creativity has also been termed as lateral thinking and represents a departure from a programmed or hierarchical thought process. It focuses on finding solutions rather than analyzing to find lacking elements. Creativity is a process that is adopted with innovation as the intended result. The more efficient the creative process, the more likely it is to result in an innovation.

In this chapter creativity and innovation are explained, and the creativity process, both at the individual level, and the group level, is described. Why creativity and innovation are important in organizations is explored. In addition to analyzing collective creativity, methods and tools to facilitate creative thinking organization wide are explored. Finally implications of those concepts and organizational infrastructure to foster creativity are delineated.

Creativity is difficult to manage (especially doing away with the notion that we can just have more by flipping a switch), but it can be fostered. Moreover, managers play quite different roles throughout the entire process, from fostering creativity to implementing innovations. Understanding those different roles across different stages will increase the probability of helping any organization take advantage of creative thinking and innovation.

27

DEFINITION OF CREATIVITY AND ITS COMPONENTS

The most basic definition of creativity is to bring something into being. It implies acting, making, or doing. We like to think something is creative if it is novel and provides a solution to some existing problem. More recent developments in the study of creativity highlight a crucial component to consider creativity is a social activity. This perspective is becoming more and more recognized and is relevant to our aim of fostering organizational or group level creativity.

Mihaly Csikszentmihalyi, author of *Creativity: Flow and the Psychology of Discovery and Invention*, is careful to remind us that creativity is a synergy of many sources, not just the efforts of individuals. Because creativity is the confluence of many people and many factors, creativity is most easily enhanced by changing the environment in which we operate and how we interact rather than hiring for creative talent. Creativity is a result of working with others (i.e., teamwork in organizations). The image of inspiration striking the genius as he or she toils away in a dark basement should be forever removed from collective thinking, for creativity depends on interaction between ideas, people, and environments.

Taking these notions in sum, creativity is then defined as *the synthesis of existing ideas in a unique way that is appropriate to the context, issue, or problem*. This definition has some profound implications for fostering creativity in organizations. Synthesis is a bringing together of elements. Hegel described the dialectic process of thesis clashing with antithesis (the thesis' own opposite) to create an entirely new concept. In Hegel's case, theses implied the birth of their own opposites. As such, the synthesis has always involved a sort of "middle ground" between two polar opposites.

Such a compromise is too narrow for the purposes of this book and underestimates the scale of transformation that creativity can engender. Creativity should not merely involve bringing together two opposite ideas for some compromise; it should involve being able to bring any two ideas together. In a less constraining case of organizational creativity, any existing ideas may be considered, and the manner of synthesis and relation between the ideas at the outset is of central importance.

This synthesis of existing ideas (i.e., the seedbed of creativity) in organizations implies two conclusions. First of all, any organization has everything it needs right now to start being creative. No additional materials or concepts must be produced before creativity can begin. The only task left remaining to complete is to bring those existing elements together in a unique way.

Secondly, "unique way" means the ideas being brought together often have little or no relation to each other before our actions put them together. In fact, the more disparate the ideas, the more creative the synthesis! Synthesis is unique because these elements being brought together do not seem to go together. Synthesis implies a healthy "stretch." What this means in organizations is that quite different ideas are brought to bear on a single problem, and that we would bring these particular ideas together to solve this particular problem nearly defies logic.

In a more contemporary work on creativity, this ancient idea is called *lateral thinking*. Coined by Edward de Bono, the main goal of lateral thinking is to break away from structured, programmed, or hierarchical thinking, where particular objects and ideas "belong to" certain classes. In creative pursuits, what is "supposed" to happen or what "belongs" is rejected. What is "supposed" to happen comes from the former conception of creativity. Any notions of what is "supposed" to happen will hamper creative thinking. A simple and common example is putting a shovel and a knife together in order to create a cake server. Two elements, which would not normally go together, form a unique solution to a particular problem—keeping pieces of cake whole while cutting and serving.

Creativity also means doing more with what we already have on hand. Though the creative process is usually problem driven, creativity as defined here is quite a departure from traditional problem solving modes. In traditional problem solving modes, the problem is defined, and what is needed to solve the problem is assessed by performing a needs or materials assessment. After examining the inventory of materials on hand, what is needed is determined, but the problem is not necessarily solved. This type of analysis is helpful and does have its role in organizational management. In fact, this traditional approach underlies most of strategic management and organizational development.

A creative approach, however, is quite different and requires a different mode of thought. Coming upon a problem, the materials on hand are examined to see how they might be used to solve the problem. No time is spent pointing out what is lacking to do the job. In other words, rather than embracing this "mode of lack," we engage in a mode of possibilities.

For example, reengineering will commonly show some deficit, identifying what is lacking but needed to do the job. While this analysis can be helpful, it is not part of any creative pursuit. Creativity seeks to explore how the job can be successfully completed with whatever is on hand. The comparison logic may look like this:

> Reengineering Questions: Here's what we've got.
> Here's what we have to do.
> Here's what else we need to do it.
>
> Creativity Questions: Here's what we have to do.
> How do we do it with what we've got?

In a creative mode, how to get the most out of the resources at hand, rather than identifying what is lacking to do the job, is stressed. Instead of fretting over what is lacking, the task is completed with the materials on hand. In this mode, materials often change their accepted purpose in order to meet the task. Creativity is a solutions oriented action in almost direct opposition to a needs oriented analysis.

WHY CREATIVITY IS ESSENTIAL

Creativity is important for the long term survival of companies and organizations. Knowing that long term well being depends on the ability to re create both ourselves and organizations as environments and technologies change, this definition of creativity becomes essential to survival. Too often creativity is treated as a luxury and performed only when time permits. In the knowledge age, however, if time is not spent pursuing creativity, organizations will not survive. Letting a company collapse is an awful way to free up one's calendar.

Creativity is a prerequisite for innovation. In other words, creativity is a necessary but not sufficient condition for innovation. Innovation requires creativity, but creativity does not necessarily lead to innovation. Management does have a role to play in taking advantage of creative thinking by promoting innovation.

Creativity is part of continuous improvement. While organizations can use practices found elsewhere to constantly and dramatically improve their current products and processes, creativity can encourage greater leaps and establish new starting points or products to improve. Whatever our current idea, design or product, ideas can always be combined and brought to bear on existing ideas, services, designs or products to make them better. Creativity itself can become a continuous process that enables ongoing innovation.

Other benefits of having a creative organization include speed and flexibility, both in design and problem solving. Today's complex environments require organizations to be more nimble and flexible. Creative organizations are quick to sense and react to changes in those complex environments. They can also be proactive and become the catalysts for changes, leading their industries in new practices.

SOME CREATIVITY MYTHS

Two things must be kept in mind when exploring what it takes to be creative. First of all, creativity is not a special talent, and secondly, creativity can be practiced by anyone. Creativity is no longer thought of as only a special talent or serendipitous outcome, but as a collective process which can be fostered. Popular images of creativity come from stories such as Thomas Edison stumbling upon the idea of using thread as filament for the light bulb, or Alexander Fleming's discovery of penicillin when he found that a mold in a culture dish had ruined one of his experiments. These powerful stories have fueled common myths about creativity being largely accidental and happening to "creative types" who see things differently.

Edison, however, did not stumble upon thread; he tried thousands of filaments and had a team of people searching the world over for better filament materials. Fleming's penicillin was not elaborated until he worked and interacted with several scientists who helped explain and then capitalize on the mold occurrence. Certainly both Edison and Fleming were incredible individuals, both known for their determination and knowledge of their respective fields. These so called "accidental success" stories cause us to overlook the role that their knowledge and determination played in their success.

Another myth promulgating mantra of creativity is the call to "think outside the box." Upon closer inspection, the underlying assumption here is still that something wonderful and unpredictable will happen if appeals are made "outside of us." Such thinking asserts that inspiration will strike "from outside" if we open ourselves to the possibility. No one can deny that it is certainly helpful to be open to outside influence and to new experiences and perspectives. Another positive aspect of "thinking outside the box" is that it encourages a focus on process instead of divine creative talent.

While this reasoning may partly be true, such advice can misdirect efforts at being creative. Such perspectives are part of what it takes to foster creativity, but the call to think outside the box can be perplexing. Thinking outside the box is basically a well intentioned call for lateral thinking, but it also provokes imagery of ideas or inspiration coming from somewhere mysterious. While thinking outside the box is a wonderful idea, it is difficult for us to summon from within ourselves thoughts that we do not have. Encountering, and then considering, diverse ideas other than those already contained in one's "box" is important, but such an exercise requires action and interaction as opposed to some ability a person has to "think thoughts he/she does not have."

FOSTERING CREATIVITY

Instead of "outside the box" thinking, the image of "shaking up your snow globe[1]" is preferable when describing the creative process. Shaking involves action and comes from collisions with other people's ideas or from experiences that require us to reshuffle our ideas in order to make sense of events or objects. Mixing up the contents of the box so that they fall together in new ways allows us to see in new ways; in other words, thought patterns are rearranged. In addition to taking action, shaking also recognizes that the process can be uncomfortable.

No matter what the problem is or how intractable it may seem, I commonly advise people to put it in the back of their minds as they stroll through a museum. In doing so, you are not just hoping inspiration strikes; you are taking some action and engaging in an experience that may cause it. Look at the art, pieces or exhibits in a museum and entertain the idea that each holds a piece of the solution to your problem. How is each image or installation like your problem?

Reshuffle your thought patterns to make sense of the objects. Your internal thought patterns—not the external experience—are being arranged here. You will begin to see your problem in a new light provided by the reordering of your ideas. You will also be out of your office, on your feet and in a new physical setting. These factors all help in approaching the problem from a new angle.

Being open and focused on processes rather than on outcomes are essential throughout such an activity. In taking these actions and "giving our own snow globes a bit of a shake," problems may be approached from a different angle. Ideas not only are rearranged and fall together in new patterns, but new ideas are also derived from the experience. The problem can now be approached from a much newer and better place. Such actions can really help to foster creative insight.

---

[1]I owe this imagery to Joe Keefe, one of the founders of Second City Communications in Chicago, who conducts creativity training and workshops based on improvisation techniques developed at the world famous Second City Theater. Joe used this image to describe his actions in getting clients to think laterally.

Organizations can work toward increasing organizational knowledge as well as encouraging and allowing teams to be determined in approaching tasks (as opposed to being cautious). Instead of hoping they are gifted with creative types, organizations can engage in creative processes and take creative, fostering actions. Companies such as Hewlett Packard, W.L. Gore & Associates and 3M are known for being consistently creative and innovative. These organizations constantly practice creativity and reap the benefits of innovation.

Nothing is innately different about these companies other than the tactics they implement and the practices they foster. 3M, known as the innovation company, has a culture of creativity and has organized its work to allow creativity to happen as an everyday practice. As a result of these creative processes and managing the resulting innovation, 3M is one of the two most innovative companies in the world. Any organization can make similar decisions, practice them, and thus increase the probability of consistent creativity and innovation.

WHAT IT TAKES TO BE CREATIVE

We have already debunked some myths about creativity being an individual talent and discussed characteristics such as knowledge of a field, social interaction, openness, willingness to take risks, and divergent thinking. Focusing on actions that rely on these characteristics of creativity is important when discussing what it takes to be creative. The impetus for creativity is the redefinition of our ideas or the exposure of those ideas to other ideas. Both redefinition and new exposure rely on having some type of experience occur, so it is an active process.

STUDY

Knowledge of the field is the ante, or the price of entry, into the creativity game. Without this domain knowledge, no indication exists of what is improving or in what ways improvement is happening. How can one expect to revolutionize space travel unless he/she knows a great deal about aerospace engineering? In the rare cases where novices do make some great creative contribution, that contribution is always as a result of working with people who are familiar with the field. In such a case, an idea from one domain is transferred into another domain where it makes an original contribution (similar to a creative collision below). Such a case reaffirms that creativity is a social, not an individual, pursuit. For organizations, the message is that if domain knowledge is not present, employees can either study to become knowledgeable in the subject, or they can work with people who do have that expertise.

COLLISIONS

While we like to think "accidents happen" in a positive and serendipitous way, in fostering organizational creativity, we have to become accident makers. Fostering these creative collisions is one way managers can encourage creativity at work. Ideas are not going to collide into anything unless they are intentionally bounced off of each other. Snow globes cannot shake themselves.

Bringing more and diverse knowledge, skills and abilities to the table encourages many ideas for solving a problem. In creative synthesis, more ideas mean more material to bring together in greater and greater combinations. Some of these combinations will unlock the problem, and the greater the number of combinations, the greater the chance of unlocking the problem.

Collisions can be facilitated by drawing analogies and generating metaphors. Using analogies and metaphors to describe similarities between two ideas (or products or services) reveals new aspects about each idea. Describing how a particular product is a lot like sailing, for example, and forcing connections between these two concepts will allow the product to be conceived of in a different way. Making these forced comparisons aids in the creative collision of ideas and helps with group processes in directing collective efforts. Different people sometimes have different ideas about a particular product and/or service that are revealed through comparison to something else.

Another way to cause collisions is to try the counterintuitive approach. In retrospect, creative solutions always seem counterintuitive. Entertaining counterintuitive ideas and actions will therefore foster creativity. Thinking in a counterintuitive manner will often spark new thinking, even if the attempts are not successful. This new thinking is another result of engaging in some strange experience.

ShoreBank is an example of a company engaging in counterintuitive thinking—even in the highly rational financial and banking industry. ShoreBank focuses on communities where other banks would not dare to tread. ShoreBank was founded in 1973, where racial and income discrimination in the banking industry was rampant. ShoreBank focuses on providing loans to entrepreneurs who play a role in improving the underprivileged local communities they serve. As a result, these communities are being revitalized, and ShoreBank is making a real difference. Where other financial institutions would consider these areas and client base too much of a risk, ShoreBank makes more character based decisions, counterintuitively seeking out communities that other banks avoid. This approach has improved local communities as well as strengthened ShoreBank's bottom line and its own clients' financial and social capital.

BRAVERY

A risk is always taken by engaging in creative processes. In this respect, creativity is a real act of bravery. In business, before time or money is invested in anything, some return on the investment must be considered highly probable. Very often these returns can be quantified and even used to help decide where future time, effort and money should be spent.

Creative processes do not lend themselves to such analysis. The outcomes cannot be predicted on a cost return basis. Because of this unpredictability, creative outcomes are always thought to be "unexpected." In reality, these outcomes were unpredicted—at least not by the usual cost/benefit analysis. While the effects of innovation in terms of cost savings can be measured, this is a matter of hindsight—not foresight.

Creativity is an exploratory process, so a willingness "to not know the outcome and yet proceed" is required.

TOOLS

Most of the materials, objects and ideas that are used are pre-defined. Hammers are for hammering nails. However, as the familiar saying goes, "When the only tool you have is a hammer, all your problems begin to look like nails." While people have a compulsion to redefine the problems they encounter to match their tools or skills, another available option is to redefine the tools themselves (i.e., a more creative option). Changing problems to match the available tools, skills and abilities (or worse—simply ignoring problems when the "proper" tools are not on hand), while both natural and comfortable, is also counterproductive to the creative process.

When the purposes and scope of available skills and abilities are redefined, more and more may be completed with the tools on hand. Remember that a basic approach to creativity is doing more with what is already on hand. Redefining tools is much harder than redefining problems, as redefining tools can be threatening. If redefining tools rather than problems becomes the norm, however, solutions will come much more easily. Additionally, the more that the skill of redefining tools is practiced, the better that skill becomes. Soon, no problem will exist in which knowledge, skills and abilities are not considered in order to solve it. Finally, the "real" problem—not the way we want to see the problem—will be solved.

Because redefinition is a process, engaging in it anytime is possible within organizations. "Creative" can become the normal mode of operation. No longer is creativity something that strikes willy nilly; activities that foster creativity can be undertaken. The assumption that creative people do things differently is partly true. Creative people learn to be open to redefining roles and tools to meet a task instead of redefining the task to match the intended use for the tools they have.

W.L. Gore and Associates goes as far as it can to avoid definitions. While many creative companies have job descriptions that are flexible, Gore & Associates does not have job descriptions at all. People are attracted to projects and then determine how their skills might make a contribution. Such a culture brings a diverse set of committed people together in a way that traditional human resource planning could never predict. Teams organize around opportunities, and leaders emerge in what is described as a flat lattice organization. No predetermined chain of command found in traditional hierarchies is present; instead, Gore and Associates employees communicate directly with each other as required by projects. They foster an environment that combines freedom within a culture of diversity that acts as a catalyst for participation, communication and creativity.

EXPERIENCE

This action not only refers to experience found on resumes, but also to engaging in activities that force a reconsideration of the accepted way of performing tasks. It involves seeking out new experience that helps drive processes such as creative collisions. Jay Conger refers to creative leaders as information collectors with a difference. He notes that they pay close attention to whatever is going on and actively seek out multiple and unrelated sources of information. This type of diversity in information seeking breeds creativity. The more "far out" or seemingly uncomplimentary the ideas are put together, the more creativity is fostered.

Not only are information seekers needed, but the importance of having experience seekers must be stressed. In addition to seeking out the diverse experience of others, we should also seek out, and take part in, diverse experiences in and of themselves. Making sense of those experiences requires one to change existing materials.

Traveling internationally forces one to realize that current ideas and customs are sometimes useless in making sense of foreign practices, especially if those ideas are applied abroad just like they are at home. Traveling in a foreign country forces a person to experience a new environment. A person either has to become open to other ways and change his or her own, or worse, that person might refuse to interact with foreigners to avoid conflict. Sadly, people who choose this route also avoid learning new things.

Most companies wait for the environment around them to change before they change. Why not seek out new environments to proactively spark change? The experience gained in learning to operate in new environments long before they reach the rest of the industry is a potential benefit of this approach. When that happens, competitors will find themselves in a strange place, but it will be familiar to us. Adaptation skills are learned, and those abilities will give a competitive advantage.

IMPROVISATION

In improvisation, acting and thinking are simultaneous. Improvisation happens, however, not because we are on "automatic pilot" in a routine situation. Improvisation involves complex, nonprogrammed decision making where design and execution are simultaneous. In other words, it involves two major components: "making do" and speed. Making do also shares many overtones with creative syntheses and redefining available tools. Making do involves putting occasional far flung ideas together in order to solve a problem or accomplish a task. Making do allows for rapid response, and more and more environments are requiring this type of action. If companies have to keep up in ever changing environments, then improvisation is a wonderful practice.

Improvisation exemplifies our definition of creativity. In improvisation, tools, not the problem or task, are redefined, and materials are used for "unintended" purposes. In organizational improvisation, skills and abilities are brought together in a solutions oriented mode (i.e., a type of creative collision). In improvisation, more can be done with whatever is available.

To illustrate improvisation, in one creativity workshop, about a dozen executives were split into two groups. Each group was given a model car—the type you might find in any hobby shop or toy store. The particular model had about 50 individual pieces and had a picture on the cover of the box showing what was "supposed" to be built with the materials inside. Instead of telling them to assemble a toy car, the two groups were told to assemble a coffee maker. Their respective experiences demonstrate what it takes to be creative.

One of the groups did not fare well. They could not get the model car image out of their heads and were extremely frustrated by the task. They exclaimed, "These are all car parts! How are we supposed to make a coffee maker out of this?" They tried for quite a while but could never get over the fact that they did not have the material they thought they needed. In the end, they produced nothing and claimed they had material unsuited to the task.

The second group was incredible. They laid the materials out and focused on the task, excited by the challenge. They seemed to think, "Well, this isn't the best situation, but we think we can try to make some kind of coffee pot out of all this stuff." In this group the "car parts" were already beginning to be redefined. Even though the car parts were initially redefined as "stuff," this small and simple redefinition was profound. It opened up a world of possibility.

The other group never had anything but "car parts" and could not detach from this confining view of the materials. All they were ever going to be able to make with their "car parts" was a car. The second group, though, could make anything with their "stuff," and this group also improvised a lot. These executives began arranging and setting up pieces without exactly knowing what they were doing. They were acting before thinking, and seeing what they were doing led to a lot of ideas.

What that second group made and described was a unique coffee maker that acted as a "reverse" French press, where water and steam were pressurized and "pushed up" though coffee beans and a filter into a receptacle. They surmised that letting water slowly rise through the coffee grinds and filter instead of quickly fall in a downward flow through a filter and into a decanter would make for a richer brew. They hypothesized that infusing water up through the beans would result in much less oily product and less coffee grind sediment than traditional drip methods. Members of that group said the impetus was what had been the car's radiator. That creative collision, coupled with materials not intended for the purpose being redefined, led to a genuinely creative idea! Will the coffee pot work? People with more domain knowledge must make that decision.

Why was the second group open, able to redefine the tools and improvise while the first group could not? The answer to this question can be found by examining each group's processes. One group did not have any more or less "creative types" than the other. What can be said with certainty is that the second team's abilities to redefine materials, seek creative collisions and improvise were big factors in fostering its creative process and the resulting success. This group also had a lot of openness and bravery combined with a willingness to try, to take a risk and to put ideas together in an uncharacteristic way. As Henry Ford quipped, "Whether you believe you can, or whether you believe you can't, you're absolutely right."

PLAY

All of the aforementioned actions require play. Creativity requires the play of ideas. Play is also the ultimate form of learning. As children play house, they learn family social structures (both good and bad). We play at physical games until our bodies become adept at muscular coordination. At work, we might shadow mentors during a training period. During play we learn to interact and perform certain tasks by mimicking actions we observe. To play means "to act in a particular manner."

Play means "to take part in," as in taking part in determining the direction of an organization in participative management. We all take part in making contributions to our organizations. To play means "to engage in something that entertains." Creative processes are fun. While it may be uncomfortable (as roller coasters can be), we seek this type of stimulation. To play is "to perform," as in a musical instrument, but this idea may also be applied to organizational performance. Play in corporations is often seen as something done when one is *not* performing. Instead, playing is performing, and it is the only way to get better at something.

To play is to compete or to try at some task. Ideas can compete, and tasks can be attempted even if they are thought to be implausible at first. This notion is related to the next definition of "play." To play is "to act in jest." Creative attempts do not have to have serious implications or be tied to success or failure at all. A low risk environment should be fostered—one where failure is not devastation but is instead a learning experience that is encouraged—no matter what the outcome. To play is "to gamble." Creativity is a risky proposition, as it could wind up changing the way everything is done. That prospect can be terrifying.

Finally, to play is "to pretend to be" and is the seed of complete organizational transformation. "Pretending to be" removes all limits and opens a world of possibility. In pretending, the corporate mission can be fulfilled in the imagination. Not only could a more aspirational mission result, but the actions needed to get there become more concrete and clear. Play allows increased clarity in creative processes.

EVE OF DESTRUCTION?

The actions described above are quite radical for most organizations. Recommending collisions, making messes, wrecking people's tools, accomplishing tasks "the wrong way" on purpose, and playing around at work are extremely atypical activities. Such activities seem threatening, and they can be, but they are also very rewarding. Much of the research on creativity is inconclusive, though the idea that creative companies survive and prosper is a proven one.

One comforting thing research has shown is that simply trying to be creative results in more creativity. Simply engaging in the activities described above fosters creativity. So perhaps the only wrong way to be creative is not to try at all. Learning is often uncomfortable until the new process or concept is mastered. We pursue challenges to make ourselves better. Remember, the goal is to shake snow globes—not break them. Shaking them can be rattling, but it is also the only way to get the pretty scenery.

These methods are anything but efficient. Efficiency is a primary aim of traditional scientific management. However, creative processes by definition fly in the face of convention.

For example, efficient information searches look to collect data quickly and from as few sources as possible, and efficient operations have a specific tool for every task. In creative pursuits, efficiency is of little help and may be a hindrance. Efficiency certainly has a place in organizations; it just does not have a place in the creative process.

Other inherent tensions are present in the creative process. One example is the tension between creativity and consistency. Consistency in the way operations occur is expected, and we constantly face tension between the simultaneous need to recreate and rely on the way we operate. Tension also exists between freedom and control in organizations as well as between individual freedoms versus the organizational values. This tension lessens when the effort is on collective or organizational creativity. Collective creativity involves the enactment of shared values in achieving some organizational goal. The effort focuses on production and improvement. Efforts to control almost always stifle creativity.

CREATIVITY TO INNOVATION

Creativity is a process in which innovation is the outcome, but creativity and innovation are highly interrelated. Creativity is the prerequisite for innovation and organizational transformation, because without creativity, innovation has no content, and transformation will be little more than more of the same. Without innovation, creative ideas are never applied. Creativity creates a solution to a problem, and innovation involves implementation.

Creativity and innovation might be thought about in two stages. The creative stage is "far out" thinking and violates current assumptions, usually generating some new ideal. Innovation moves towards that ideal using analytical thinking. Scientific methods are relied upon to bring the creative solution into reality by determining the best way to re fabricate current practices, policies and structures to coincide with the new way of thinking. The creative reorganization becomes the new standard through successful innovation.

In innovation, many of the same organizational challenges encountered in creative processes are present, and management plays a vital role. Remaining open and having a highly supportive and participative environment is crucial. Autocratic styles of leadership, rigid bureaucracy and/or highly functional structures can hinder innovation. In addition to supporting risk taking, managers must provide reward systems to support adoption of the innovation. If implementation is of extremely high priority, managers will need to be proactive about the type of pressures employees face.

Any pressure should be oriented around achievement and performance. Different from the stress of time pressures, the pressure to perform is based on goal setting and relies on inspirational leadership. Organizations will need open communication channels and free flow of information. Organizational designs should be organic and adaptive, as mechanistic and rigid structures inhibit innovation. Finally, sufficient resources must be provided in order to make innovation successful.

Some distinct challenges to innovation exist. Innovation or creative application results in paradigm change, which is completely reconstructing the way current processes are seen and performed. Some profound implications result when the "new" paradigm is put in place (and the "old" paradigm is retired).

First, understanding the new paradigm in the language of the old paradigm is virtually impossible. Therefore, no amount of talk, verbal demonstration or logical argument (no matter how eloquent, emotional or rational) can convince members of the old paradigm that the new paradigm is better. Any argument extolling the new paradigm is made from the perspective of the new paradigm—a perspective simply not shared by members of the old paradigm. In addition, members of the old paradigm will have a solid, rational answer for why every single innovation "won't work," which will be air tight logic in the language of the old paradigm.

In such a situation, action and trust must be the hallmark of the new paradigm's introduction and implementation. Talk simply will not suffice; in fact, the two sides do not even speak the same language. The new paradigm is realized as a result of taking action. Innovation must be enacted into reality, and explanation will follow.

Just as a willingness to take risk and a supportive environment allow creative thinking, the same can be used for implementing innovation. Once the new way of thinking is brought into being, the evidence of the innovation will be undeniable. The old paradigm is also used to explain the events. When the old way of seeing and thinking cannot explain the events, and the new paradigm can, the new paradigm is adopted as well as the innovation that results.

Two additional practical issues are encountered during innovation. One is that many ideas are not so creative after all. This fact may not become evident until the practicalities of innovating are in place. The lesson here is that creativity is a practice, and going back to the drawing board is possible at any time. The more that creative processes are practiced, the more creativity that results. In the meantime, small improvements can be realized.

The approach to creativity and innovation must always be one of possibility, where questions revolve around how to make an idea work—not listing reasons why it cannot work. The "why it won't work" list is always easy to generate. In attempting to make a creative idea work, we may only get half way to our ideal situation. In such a case, should we consider this innovation a failure? Not by any means. In only getting half way to our ideal world, are we still not a lot better off than where we were before? And what is the alternative—dismissing the creative idea completely because of doubts about the ability to innovate? Rejoicing in the success that is achieved, and seeking continuous improvement in this creative process in order to become better innovators, should be the goals.

PRACTICAL EXERCISE

Here is a creative exercise that demonstrates the concepts presented in this chapter. It is also an exercise that can be applied in most any situation in any organization. Breaking perceptual sets (the way we see) and cognitive sets (the way we think) can aid in arriving in a creative mode. Six steps are necessary to improving any existing idea, product or service with creative thinking.

STEP 1: Assemble a diverse group of people for this exercise. Beyond people from every level and across departments in the organization, it helps to include people (including competitors) from outside of the organization. It helps to have a complete stranger in the mix as well.

STEP 2: Think of the product, service or idea that needs improvement (called product X in this exercise).

STEP 3: Use brainstorming techniques, such as suspending judgment, and have everyone contribute as many ideas as possible; list all of the attributes that make for a great "product X" in general.

STEP 4: Looking at those listed attributes, think of some OTHER product/service/idea that has similar attributes. List a few of those and pick one that resonates with the group (but is not too similar). Having too similar of a product will not make for much of a creative collision.

STEP 5: Now list the attributes that result in a great "OTHER product" in general. The real challenge here is in forgetting what was listed in step 1. The focus must be on a great OTHER product (step 3). Keep ideas focused on improving this OTHER product/service/idea. (I always cover up any product X materials.)

STEP 6: Taking the list from step 5, brainstorm how those ideas "translate" back into "product X." Be very open here. Think of how such a product can work. Entertain the craziest of ideas, even if they feel forced. How would something listed for OTHER product work or look like if that attribute was a part of product X? How can one use an attribute from the OTHER product to improve product X?

After step 6, reflect on the ideas that are most novel, appropriate and satisfying and consider implementing them. Re create (i.e., innovate) your product/service/idea.

CASES

This creative infusion exercise has happened many times, and I am always amazed with the results. In one simple case that highlights a product improvement, I had a group of students thinking of ways to improve cell phones. Their OTHER product was a personal assistant. One of the attributes of cell phones is that people are "always in touch." One of the attributes of a personal assistant is that people using them always knew where they are and where they are supposed to go next.

Out of this creative collision came the idea that GPS locator chips should be sold with cell phones. The group members recommended using GPS tracking and interactive map technology (bundled with the cell phone) that works with button sized "locator chips." The location of the "buttons" would always be shown in relation to the cell phone. For example, you could leave a button in your car on your way into a crowded stadium and then use your cell phone display to locate your car after the game.

In shopping malls or amusement parks, parents could sew a button onto a child's jacket or into a child's pocket, so even when the child is out of sight, he or she could quickly be located via the cell phone. The display would show the location of the cell phone and the location of the button(s). In a more ominous scenario for business travelers, you could place locator buttons in your luggage as you travel. If the distance between you and your bags is increasing rapidly as your plane takes off, at least you will know your luggage is lost before the airlines do!

In a case that relates to service, I recently conducted a workshop at DePaul University's Center for Creativity and led a group through a creative infusion to improve an executive search firm. The participants in that workshop chose a professional sport franchise as their OTHER product. Some of the ideas and improvements involved language training, relocation and the promise of jobs for spouses and schools for children. Those services were new for this firm, but competitors already offered these services to their clients. The participants also recommended a change in perspective, acting more as agents for the executives placed in addition to providing services to their client firms where executives were placed.

Though the corporations were paying the bills, the search firm knew that its reputation also depended on the personal experiences of those placed in positions. The firm's goal was to become the executives,' as well as the companies,' search firm of choice. This firm felt confident it provided a superior service to corporations, so any improvement could only enhance the experience of the executives it placed. Such a change in perspective has a lot of potential.

Creativity can start to be fostered by taking action and engaging in these creative processes and practices. None of them require people to be "creative types." Though the processes may not be conventional, the basic aim is one that resonates with organizational goals. For instance, doing more with what is currently available is creativity's approach to organizational efficiency. Being flexible and improvisational are excellent organizational design components and necessary for organizational survival. Lastly, these processes are ones which can be mastered the more they are practiced. In no small sense, learning how to be creative involves learning how to learn.

TAKE AWAY

1. Creativity is not a special talent and can be practiced by anyone.
2. Creativity is a synergy of many sources—not just the efforts of isolated individuals.
3. Creativity depends on interaction between ideas, people, and environments.
4. Long term well being depends on the ability to recreate ourselves and organizations as environments and technologies change.

5. Common myths about creativity being largely accidental and only occurring to "creative types" who see things differently do exist.

6. Collisions, making messes, wrecking people's tools, doing things the wrong way on purpose, and playing around at work all seems threatening, but these activities can also be very rewarding.

7. In the rare cases where novices do make some great creative contribution, it is always as a result of working with people who are familiar with the field of study.

8. We are very well practiced at redefining problems to suit us. Redefining tools, however, can be threatening. Admitting our skills cannot solve some problems is uncomfortable and not easy.

9. Creativity is the prerequisite for innovation and organizational transformation.

10. Creativity creates a solution to a problem, and innovation involves the implementation of that solution to the problem.

11. Understanding a new paradigm in the language of the old paradigm is impossible.

12. We must still be vigilant that our approach is one of possibility, where questions revolve around how to make an idea work—not listing reasons why it cannot work.

THE CONVENTIONAL TOOLS OF CREATIVITY

Jim Harrington

In business, creativity is used to solve problems. Therefore, the greater the amount of creativity that can be unleashed, the better the solution is likely to be. As a result, over 200 tools have been developed to enhance and extract the creative talent in individuals and groups. The effectiveness of these tools is a function of people, the environment, process need(s) and the problem being addressed. These tools help in developing creative focus, setting direction, exercising the mind, shattering paradigms and gaining new insights. These tools assist in creating a framework that over time has consistently yielded better results.

Humans are tool-making creatures, and much advancement can be traced to humans' tool-building abilities. We need tools because our senses of touch, taste, smell, hearing and seeing have a limited range within which they function. But even in these functional abilities, living things do exist that can beat people hands-down in every sensing category—a cheetah in speed, an eagle in eyesight, an Arctic tern in navigation, a moth in smelling, etc. The only tool in which we seem to excel is "minding"—the ability to comprehend, memorize and think. The tools of innovation help us in "minding" and provide various methodologies by which we can comprehend where we are, understand where we want to go, know what we do not know, and think creatively how to get there.

Any sensory information lying outside the ranges of our senses needs to be converted and brought into our range for analysis. Thus we have arrived at much of the "magic" of technologies around us. Also, our day-to-day experiences show that our sense perceptions can be misleading. Sense perception is self deception—thus the need for "minding," or the inquiry and analysis to get at the truth. Irrespective of the input to our senses, inquiry alone brings out the truth.

We look at innovation tools as necessary but not sufficient requirements for innovation to occur. Over 200 tools are currently available and more are being "innovated." Becoming familiar with these tools, learning how to use them and anticipating what results can be obtained are all important steps in the creative process. Gaining practice in using selective tools for obtaining mastery of their use and excellence in results is essential. The caution is also in using the right tools for a given purpose. As the adage goes, "With hammer in hand, all problems appear as nails!"

The serious innovator identifies the innovation strategy for the company in terms of Design Strategy, Consumer Centric Innovation, or any of the other approaches used to become familiar with the needs of the business. After all, innovation is the creative application of our imaginations to *meet the needs of the business*. The same applies to maverick innovators engaged in identifying and meeting the needs of society.

BECOMING CREATIVE

The cycle time of ideas, from input to creation or capture of output, is very small. The cycle time for products, processes and businesses based on ideas, however, is much larger and thus slower. The economic pressures are such that the need for changes, especially innovative changes yielding specific benefits in cost, speed, quality and new products/services, is accelerating and creating demand for new innovation tools.

Branding, differentiators and innovation are gaining focus in all business activities. In meeting the needs of the market, a business cannot grow and profit as a "copy-cat." The business needs to establish its own identity by excelling in one or more areas and being well ahead of the competition in many others.

CAUSE AND EFFECT

Understanding the cause-effect relationship is essential in running a business and managing involvement in its differing phases. Well-established rules, relationships and policies ensure the continuity of a business organization. When creativity is considered, however, that very same knowledge binds one from being successful in these efforts. Phrases such as "thinking out of the box" help identify and remove the boundaries that impede creative thinking. Being able to recognize boundaries is a much-needed skill both for successful management, which tries to keep the boundaries, and also for innovators, who need to stretch, shift or break these boundaries.

THE OSCAR AND FELIX APPROACH TO CREATIVITY

We all have two personalities that live inside of us—much like Oscar and Felix of "The Odd Couple" who live together. Felix "occupies" the left hemisphere of the brain, and Oscar "occupies" the right hemisphere of the brain. Felix is organized, highly literate, a master list-maker and planner, and he never deviates from the plan. He is driven by rules and the clock. He develops new rules, strives to please others and is very disappointed if others do not recognize his efforts.

Oscar's personality is unstructured, reactive, and driven by whims. He drinks beer for breakfast that was left opened the night before. He challenges authority and rejects conformity. He feels best when he is working on many items all at the same time. He believes that rules are made to be broken. He works to his own beat and relies on self-gratification to keep him going.

Felix had a 4.0 average in college. He loves exams because it proves to his teacher that he did the assignments and learned his lessons. Oscar had a 2.0 average. He created problems in class. He told jokes. He was more interested in making friends than in making grades. While Felix works to accomplish something, Oscar functions for the joy of doing the task.

Education, a structured work life, and the need for conformity have dulled creative pursuits. Many of us have become like Felix. The Oscars among us, or those with a rebelling Oscar in them, tend to break the mold and venture outside the boundaries in order to create something new.

LETTING GO: WAKING UP OSCAR

The only person who can make someone more creative is that someone! If the reactive life has been draining, explore the proactive approach. Such a move may require some preparation. Create a conducive workplace, release tension by any techniques available, relax and meditate.

- *Time.*   Extra time is often required to develop and sell a creative solution that is not in line with the organization's culture.
- *Environment.*   Being continuously interrupted with phone calls, questions or family demands does not contribute to creative pursuits.
- *Success.*   Being recognized for conceiving of, or contributing to, creative solutions really gets the attention of a "Felix."

STRUCTURED TOOLS OF CREATIVITY

All tools of creativity have produced results, and many users vouch for their effectiveness. What tools, however, will work for you? A photographic plate works only when it is prepared and ready to change when the image is reflected onto it. In current terms, digital photography only exists when someone can store digital images onto a drive with its accompanying changes in the drive. By the same token, in attempting to use the tools of innovation, being receptive to changes that the tools will demand is critical.

No change in ourselves—no effective tools of innovation!

When one thing changes, everything changes. Hence the noticeable great power when one individual changes the world by changing him- or herself. Before any tool of creativity is adopted, this question demands an answer: Am I ready to change and in what way? What is the output from the tool that will help me with that change?

1. Developing Creative Focus

A mind that knows exactly what it wants may miss the opportunity to expand its capability. Merely seeking capability may prevent the mind from finding it or recognizing a possible opportunity. Many tools exist that are designed to expand the mind. Mind-expanding tools were not designed to solve problems or to create new concepts; they were designed to aid in unleashing creative powers. They fall into simple, medium and advanced mind-expanding categories.

Styles of Creativity

Many styles of creativity exist and can generally be classified into the following four modes:

Structured creativity: Step-by-step, detailed, complex, tool-intensive, controlled, effective for individuals and groups, and requiring little facilitation.

Non-linear creativity: Exciting, unpredictable, fast-paced, focused on quantity and not quality, promotes people involvement, and usually used in groups.

Provoked creativity: Catalyst-focused, provides a springboard for forward movement, easy to build on, easy to start, requires active facilitation, and easily used by individuals or groups.

"Aha" creativity: Can be described as having no steps or patterns, focused on big issues, invariably has a defining moment, uses simple methods, and is individually intense.

2. Exercising Your Mind

Simple Mind Expanders

Simple mind-expanding techniques include concentration exercises. Marking the "mind beats" (by placing a pencil and making a line with full concentration at the pencil point where it becomes a line) is another technique. Each time the mind wanders off, draw a mind-beat (i.e., mark the place with a kink in the line like a heartbeat). When the edge of the paper is reached, look at the complete line. How long can an unbroken line of awareness be maintained?

Another minding exercise is to count all the capitalized alphabets that contain only straight lines (or fully-closed loops). Other games with nursery rhymes and numbers—for example, saying the number of letters before saying each word of the song—are all aimed at exercising one's mind to change the existing patterns of thinking.

Medium Difficulty Mind Expanders

Analyzing Outrageous Ideas: The goal here is to track daily outrageous thoughts (Oscar's inputs) not put to use. At the end of the day, review the list, check marking ideas that should not have been rejected and write down why they were rejected. Select a sample (3 to 6 thoughts) from this list that are still "bad" and make a list of what is good about the ideas, what they can still accomplish, and why they were suggested in the first place. Then define how each thought can be reshaped, focusing on their good points to make them acceptable concepts.

Pictures to Drive Creativity: If a picture is worth a thousand words, it can create a thousand thoughts that can be combined into many creative ideas. Cut out interesting pictures and choose pictures that stimulate thoughts and/or ambiguities. Randomly select three or four of them and create a story that includes the items in the pictures.

Words to Drive Creativity: Write down interesting nouns on separate cards like apple, Monterey, Rio, sea, gum, tower, lighthouse, radio, etc. Pull three to five words at random from the pile and make up a story using each noun.

Differences and Similarities: Shuffle the cards from "Pictures to Drive Creativity" and pull out two cards at random. Make a list of how the two items are different. Make a second list of how they are the same. An example of how cars and houses are similar is that they both use electrical systems, and people go into both cars and houses.

Define Other Applications: Another application of this pile of cards with key nouns on them is to select a card at random. Ask how this item is normally used and what other ways the item could be used.

Creative Progress Reports: At the end of each week, prepare a progress report on a 3″ × 5″ card. Take no more than five minutes to handwrite the report. The report will be made up of four sections. On the top half of the front of the card, write what was performed during the week. On the bottom half, write what was accomplished. Then turn the card over and on line one, record how you feel about your job. On the second line, record how you feel about what you accomplished. Use the rest of the card to record two ideas that would improve the way you feel about your job and your accomplishments.

Then write a statement on what you will do next week to make things better. After you have made ten weekly reports, review the ten cards. Compile a list of all the suggested changes that would improve the way you feel about your job and your accomplishments. Select the two most important ideas and prepare a plan to correct the situations within the next four weeks. Every month, update the list and define actions to correct the two most important conditions.

Each time the weekly progress report is reviewed, evaluate whether the weekly corrective action plan was implemented. Every month, calculate the percentage of weekly corrective actions that was successfully implemented. This percentage should be plotted on a graph.

Advanced Mind Expanding Exercises

Dreaming in Color: Most people dream in black and white. If you dream in color, you are one of the exceptional individuals possessing a high potential for creativity. To change your dream pattern from black and white to color requires a little practice and is readily mastered by most people. This transformation can be conceptualized as the increased impact a movie has on someone when it is projected in Technicolor instead of black and white.

To change your dream pattern, buy a group of 8½″ × 11″ colored chips, including bright red, yellow, green, blue, purple, and orange. Before bed, intensely study one of the colored chips without interruption for five minutes. Then turn off the light and for the next five minutes try to visualize the color in your mind. Each night, repeat the cycle using a different color. After a short time, the color can be visualized without studying the colored chip ahead of time. When this occurs, go directly to the visualization part of the exercise, selecting a different color each night.

Once someone is competent at visualizing the individual colors, move to color patterns. Start with simple patterns— maybe the pattern on a favorite dress or tie. Do not be concerned if the exact color configuration cannot be visualized mentally. The object of this mind expander is not to copy color patterns but to add a color dimension to one's thinking. Most people will have their first colored dream within 30 to 45 nights. Keep up the nightly ritual until most dreams are in color. Then phase out the mind expander. Typically, results from this mind expander are achieved within 45 to 60 days.

Record the Evening's Activities: Many good ideas occur during sleep. Creativity often explodes into one's consciousness and is lost before it can be captured. We have all said to ourselves, "That was a great idea, but I can't remember what it was." As we increase our receptiveness to the right-hand side of the brain's activities, these random explosions will happen more and more frequently. They will happen when we are driving to work, giving a presentation, talking to a friend, listening to a lecture, and during sleep. Capturing these flashes of brilliance when they occur is very important, since they quickly re-trench back to one's subconscious.

Record the concept so that returning to restful sleep without worry is once again possible. Keep a pad and pencil or a small dictating scratch pad beside the bed to record great, and even not-so-great, ideas that come during the night. Do not limit this concept to the bedroom. Keep a means available to record the exploding ideas all the time.

Additionally, record what you are doing when the idea occurs. Keeping a graph that records how many creative ideas you have per week is a good idea. Set a target to double the quantity over a two-month period. Also analyze this data to determine under which environments you are most creative. Use this data to create environments that increase your creativity potential.

Mind expanders that are designed to change thought patterns have now been introduced. Breaking the established patterns you have in order to foster creative efforts is critical. The first requirement in pulling up creative abilities is to get enjoyment from being creative. Therefore, no attempt has been made here to delineate mind expanders that will help solve problems. Three important truths here are:

- It is critical to expand how one thinks.
- Thinking creatively needs to be practiced on a daily basis.
- Becoming more creative is possible if one tries.

No one can force another person to become more creative; one has to desire to *practice* creativity. Just using the mind in a consistent way can help anyone become consistently creative. Therefore, while expanding one's mind is possible, doing so takes practice.

3. SETTING DIRECTION: UNDERSTANDING THE CONTEXT

Individuals do not suffer from a lack of creative ideas. Neither do organizations suffer from a lack of creative ideas. Both individuals and organizations suffer from the lack of successful application of those creative ideas. Putting ideas into action is not easy. The success of a creative idea lies in its application. A creative idea is bounded by its frame of reference.

A unique set of methods is delineated in this chapter that helps in answering the question, "Are we being creative within our frame of reference?" The four methods presented include the following: Manager-to-Manager Event, Possibility Generator, Exaggerate Objectives, and the Selection Window. Each method utilizes a different type of creativity—structured, non-linear, provoked, or "aha."

Manager-to-Manager Event

Setting one's direction requires the ability to look into the future and determine a vector—a course of action that will get a person where he or she needs to go. A fascinating method for determining and communicating that direction is a manager-to-manager presentation.

Possibility Generator

The Possibility Generator, a technique for creating a large number of new ideas in a very short time, is also examined. The Possibility Generator concentrates on the quantity—not the quality—of new ideas, thus freeing minds to create relationships between those ideas at a very rapid rate. A fixed objective allows everyone in the organization to see the organization's direction.

Exaggerate Objectives

By exaggerating objectives, or overstating the position, individuals often are able to break through the cultural barriers that they unconsciously erect. Exaggeration provides different perspectives, enabling a creative approach to an opportunity. Stretch each criterion in some fashion. Use the exaggerated criteria to stimulate creative ideas.

Example of Exaggerated Objectives

STEP 1. Define the opportunity.

STEP 2. List the criteria that will satisfy the opportunity.

STEP 3. Exaggerate the criteria.

STEP 4. Use the exaggerated criteria to generate ideas.

Exaggerating Objectives works best for groups working on issues that face enterprises. This method uses a non-linear style of creativity, and it is best applied to setting direction. The creative power of this tool is seen well using time frames of months and even years. Exaggerating Objectives provides a method to cause organizations to stretch and break out of their normal modes of thinking. This tool helps someone understand future direction.

Selection Window

The Selection Window is a technique to help someone select the right tasks to accomplish. The criteria used are effort (resources) and probability of success. Using this tool enables a person to carefully consider each option.

Begin by generating a list of options, alternatives or opportunities that can help you focus on where to apply your creativity. You may choose to use one of the tools, such as Brainstorming, Mind Mapping or Brain Writing. Numbering each option helps. Draw a window containing four equal panes or boxes. Mark the

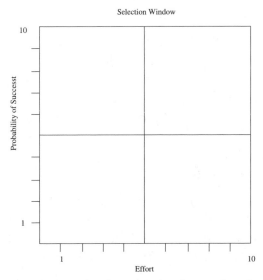

FIGURE 3.1 Plot the scores for ideas in the selection window

window with effort on the bottom and probability of success on the left. Draw a scale of 1–10 (low to high) on both the left and bottom.

Evaluate each option on the list against the criteria of effort and probability of success. Ask, "How much effort will it take to accomplish this?" and "How likely is it that this idea will succeed?" Score each question between 1 and 10. Place the number of the option at the appropriate place on the window. Now review the Selection Window and take appropriate action.

Example of Selection Window

STEP 1. Develop a list of ideas.

STEP 2. Give each idea a score for each criterion.

STEP 3. Place the ideas into the Selection Window.

STEP 4. Analyze and take appropriate action.

Pursue First: Pursue items of high probability of success and low effort first. These items are the "right" choices that have the highest probability of success. Each item could have significant impact and will incur only a small drain on the resources.

Pursue Second: Ask others for assistance with completing items of high probability and high effort. These items are important enough to deserve the attention that a skilled group of people truly working together can supply.

Pursue Third: Place items of low probability and low effort on a "To Do" list. Use these items as fillers. Although they are of low probability, they are important enough to justify a small expenditure of resources. Be careful! The line between items to be done immediately and those on the "To Do" list often gets blurred. These "To Do" items often dominate one's time to the exclusion of the most important items.

Pursue Last: Postpone items of low probability and high effort until the other options have been explored. Quite simply, the end does not justify the expenditure of the high level of resources required to complete the item at this time.

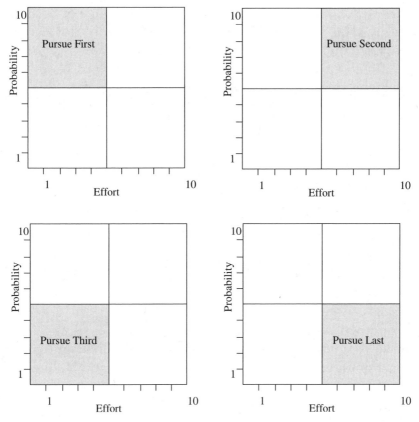

FIGURE 3.2 Analyze and Act

4. SUSPENDING RULES: GAINING NEW INSIGHTS

In many situations the normal rules of behavior must be suspended to allow more effective communication. Thus a scale for assessing communication has been devised. Listen to nearby conversations and determine where most of those comments and conversations fit. At the lowest level in the negative zone is slander, where one person tells falsehoods about someone else in order to hurt and ruin the reputation of that individual. Many individuals suffering from inferiority complexes or low self-esteem feel a need to denigrate others to make themselves feel strong and healthy.

The second level of conversation is the exchange of lies and untruths. This practice is not done to necessarily run people down. Exchanging lies is usually done just to present a distorted view of reality so the liar can make him-/herself look good.

A higher level of communication is to feel free to express opinions. Observe the following communication hierarchy:

> Needs
> Feelings
> Options
> Facts
> Clichés
> No Talk
> Misdirection
> Manipulation
> Malice

For creativity to grow, quite often the old rules of misdirection, lies, and slander need to be suspended, allowing for the free communication of open and honest needs and feelings.

Say/Think

The Say/Think methodology is a very simple form of linguistic analysis and facilitates a person's understanding of someone else.

What was said	Who said	What I thought

FIGURE 3.3 Using Say/Think

Groups formed to understand each other's position or the position of a higher authority can also use it. Using the Say/Think method can aid in personal and business relationships.

To use the Say/Think methodology, simply utilize a piece of paper with three columns. Write down what is actually said in the first column, who said it in the second column, and what you were thinking in the third column. The results may be analyzed in two ways. The easiest way is to go back after the conversation and reanalyze what was said and what you thought. A much more effective analysis is to review your sheets with the sheets of the others who were involved. If more information is needed, add a column to record what you felt whenever something was said.

Example of Say/Think

STEP 1. Draw three columns on a pad.

STEP 2. Record what is said in a conversation and who said it.

STEP 3. Write down what you were thinking after each statement.

STEP 4. Review what was said and what you thought. If you trust the person(s) with whom you had the conversation, share your written thoughts.

Say/Think is a tool that works best for groups working on personal issues. It uses a very structured style of creativity and can easily be applied to direction, planning and tasks. The creative power of this tool is best seen when it applies short time frames to provide a look behind what is actually said. This tool offers a very powerful way to manage differences and unleash one's creativity focused on consensus.

Role Play

Role Play is a method that allows a team to examine a situation from many different points of view. Using Role Play to examine a real opportunity, problem or situation is most effective. Each individual assumes the role of someone involved in the situation, such as the customer, service representative, salesperson and executive. Individuals describe the situation from their assigned perspective. Role Play can be used to help anticipate or improve opportunities, problems or situations.

Example of Role Play

STEP 1. Define the opportunity, problem or situation.

STEP 2. Define who is or should be involved.

STEP 3. Assign a role to each individual.

STEP 4. Role play.

STEP 5. Debrief the session focusing on new insights and learning.

Role Play is a tool that works best for groups working on enterprise issues. A very non-linear style of creativity is employed when using this tool, and it is best applied to tasks. The creative power of this tool is best seen when it is applied within short time frames. It provides a structured method and a look behind what is actually said. This tool provides a method to examine what is happening from different perspectives.

Cartoon Drawing

Using cartoons as a stimulus allows one to suspend the rules that dominate normal reality. The use of humor injects a different look at the way things actually work in the world. If you cannot draw, use a cartoon from some source and remove the words and titles. Give the cartoon to those involved and have them develop their own words and titles for the cartoon.

Example of Cartoon Drawing

STEP 1. Define the opportunity, problem or situation.

STEP 2. Use an existing cartoon with the words and titles removed, or draw a new cartoon figure that seems to capture something about the opportunity, problem or situation.

STEP 3. Develop new captions.

STEP 4. Debrief the session focusing on new insights and learning.

Cartoon drawing is a tool that works best for individuals working on personal issues. Cartoon drawing uses provoked creativity and is best applied to both planning and tasks. The creative power of this tool is best seen when it applies short and moderate time frames. Cartoon drawing uses a non-threatening situation to examine germane issues creatively. This tool examines what is happening from different perspectives.

Listening for Comprehension

Listening is vital to being able to communicate with others. If listening for comprehension does not happen, data are overlooked, motives are misunderstood, and distrust is promoted. To listen for comprehension, understanding must be the goal. All spoken words must be heard. The logic that the words contain must be sought. Beyond the logic, the motivation must be determined, as feelings drive motivation. When the words are understood at this level, framing the understanding in one's own words is important.

Listening for comprehension requires that listening is active, attentive, reflective, empathetic and searching. Listening for comprehension allows a person to grow in his or her understanding of others and prepares him or her to hear and understand the speaker.

Listening for Comprehension works best for individuals working on personal issues. It uses "aha" creativity and is best applied to direction. The creative power of this tool is most effectively seen as it applies to all time frames. Listening for Comprehension provides a method to examine and understand what is being said. This tool allows a person the chance to examine what those people he or she deals with understand.

5. Thinking Differently: Shattering Paradigms

A chick needs the shell to grow, but a day comes when the shell must be broken. If it is not, there will be suffering and death— Nisarga Datta

Shattering the existing seen and unseen barriers in order to break existing paradigms or create new ones is a noble and very helpful exercise.

Brainstorming

Brainstorming is a way to generate a large number of ideas in a short time. It can break existing thought patterns and generate new options. It encourages strange, wild or fanciful ideas. All ideas are valuable and no judgment or evaluation is made while ideas are being generated. Ideas can be combined to create more ideas, and the target of brainstorming is quantity of ideas (not quality).

Example of Brainstorming

STEP 1. Select a purpose for the brainstorming session. Be as specific as possible.

STEP 2. Review the rules of brainstorming (see below).

STEP 3. Generate ideas and capture them with one or more scribes.

STEP 4. Use a selection window to sort the ideas according to the evaluation criteria.

A number of variations to brainstorming are widely used. Much focus is on the pre-work or preparation made before the start of the session. Structure defines how the session is conducted. The structure for the brainstorming session may include a round robin format (where everyone gets an opportunity to speak) with a "pass" option for participants.

Rules of Brainstorming

- Defer all judgment. Do not permit criticism or evaluation. Withhold discussion and evaluation until later.
- Free-wheel. The wilder the idea, the better. A seemingly ridiculous idea may provide the trigger for the breakthrough idea.
- Seek quantity. The greater the number of ideas, the greater the likelihood is of producing a winning idea.
- Piggyback and hitchhike. Combine and improve. Build upon other ideas. Move ideas forward.

- Record all ideas on a flip chart. Use more than one scribe if needed to capture all ideas.
- Introduce competition. When possible, encourage competition between groups or challenge them to meet or exceed a quota.
- Use active listening responses. Encourage responses by smiling, nodding and making eye contact with participants.

This technique is best applied for groups working on enterprise issues. It uses non-linear creativity, which is best applicable to tasks. The creative power is most effectively seen when applied to short time frames. This tool is great for emerging groups. Keeping brainstorming sessions focused is vital.

The Six Thinking Hats

Dr. DeBono, a leading theorist of the thinking process, has introduced many effective, widely-practiced thinking tools. One notable tool is the Six Thinking Hats, where six colored metaphorical hats to represent styles of thinking is introduced. The six hats are briefly explained below:

- *White Hat:* The white thinking hat covers data, information and facts and identifies the gaps and information needs. A typical input is: "We need to step aside from our current arguments and look at the available data."
- *Red Hat:* The red thinking hat originates in intuition, feelings and emotions brought on by the current discussions. Feelings need not be justified, allowing for valuable input. This input is valuable only when grounded in logic. Normally emotions are genuine, but the logic behind them can be erroneous or flawed.
- *Black Hat:* The black thinking hat represents judgment and prudence. Though judgments are often hasty, we need to combine the suggestions as well as fit the known facts, experience, and guidelines being followed. The black hat arguments are always logical.

- *Yellow Hat:* The yellow hat stands for sunshine thinking, or bringing out the benefits of an idea and how it can work (or be made to work). This is a forward-looking thought process that focuses on results and benefits. Yellow hat thinking can also find the value of completed work as well, providing positive energy to discussions.

- *Green Hat:* Green hat thinking is rooted in creativity, innovation, alternate ways, provocations and changes. Though the energy of green hat thinking is generally positive, a certain amount of pain or negative energy is associated with it.

- *Blue Hat:* Blue hat thinking is utilized when an overview and process control are needed. This thinking is not focused on the subject but rather on the thinking about the subject. Blue hat thinking identifies "other hat thinking" needed to facilitate and make progress.

In applying this tool, facilitation training and experience may enhance the results obtained. However, any focused session to solve a specific problem can be carried out with all the guidelines applied (see below).

Suggested Hat Sequence:

- White Hat: Present information and data.
- Green Hat: Provide creative thinking.
- Yellow Hat: Support why it may work.
- Red Hat: Feedback feelings and intuition.
- Black Hat: Give rationale for why it may not work.
- Blue Hat: Provide guidance to the thinking direction.

Some Guidelines:

- Too much level disparity among members may hinder the free flow of ideas, so establishing ground rules will be helpful.
- Team members select hats, with a scribe keeping track of the output by hat or color category. Each member will rotate through each hat for a well-rounded session.

- The team follows a timed sequence of inputs from each "hat." The white hat representative usually starts with presenting the data and others follow in any sequence. Having a suggested sequence may help some teams maintain positive energy levels.
- Members change hats to the next color in sequence, which necessarily changes their thinking framework to match the hat color.
- Keeping a set time limit for each hat and the session itself is helpful.
- The team completes one "cycle of hats" and *logically supports* all input.
- The summary of all input is reviewed with filtering techniques.

This cycle can be repeated to refine or to explore more ideas. The benefit of this tool is that no team member is stereotyped or bound to a style of thinking. In fact, the tool forces each member to think in other members' "shoes" to effectively participate. This brings out the best proposals that can benefit the business.

Free Association

Free association is a method for making mental connection between two different ideas. It provides a simple way to examine things that are similar (or a pair of opposites). This technique works best for a group of people with no specific end in mind and is a very powerful tool for developing "Aha" ideas.

Free association requires linking 12 to 15 ideas before a useful idea results. Free association is a method that functions best for individuals working on personal issues. It calls on "Aha" creativity and helps with direction-setting involving long time frames. Free association is a method used to stimulate chains of ideas.

Thinking differently gives us new, rich, powerful ways to look at opportunities, problems and situations. The methods described so far help break down traditional walls or boundaries that have been created.

6. ESTABLISHING FORMATTED WORK SPACES: UNLEASHING THE VALUE OF SURROUNDINGS

Every job consists of three parts. One part is deciding on what the task is; the second part is planning how to carry out the task; and the third part is actually carrying out the task. To be creative, time and energy must be allowed for each of the topics, so a special "fertile" environment (called a formatted workspace) must be present.

Formatted thinking workspaces are created and dissolved electronically around the world via the Internet. The trend is just beginning; the surface of the power of applied technology has barely been scratched. The challenge here is not the acquisition of this technology. Rather, the challenge is found in the application of technology to creating formatted workspaces in order to allow for creativity in one's life or organization. Four methods are very helpful when working in a formatted workspace, and they are force analysis, card sort, mind mapping, and environment.

Force Analysis

Force Analysis is a technique used to make conflicting forces visible. It provides a way to visualize growth forces versus restraining forces. Force Analysis also provides a map of internal forces versus external forces. It can help you to discipline your thinking and identify the keys to success.

Example of Force Analysis

STEP 1. Define the opportunity, problem, or situation.

STEP 2. Draw a large circle.

STEP 3. Define or describe what is preventing success (restraining forces). Draw an arrow outside the circle, pointing toward the center of the circle, for each force. The length of the arrow should indicate the strength of the force that is working against success.

STEP 4. Define or describe why success is important (driving forces). Also identify anything that can encourage success. Draw an arrow inside the circle, pointing from the center of the circle. The length of the arrow should indicate how strong the driving force is. Continue the process of identifying and marking the driving and restraining forces until ideas are exhausted.

STEP 5. Analyze each of the forces and consider the following questions:

- Is there a driving force that can overcome each resisting force?
- Can the resisting forces be reduced?
- Can the driving forces be increased by strengthening them or adding new ones?

Force Analysis is a method that works best for groups working on enterprise issues. It uses a very structured style of creativity and is best applied to tasks. The creative power of this tool is best seen as it is applied within short time frames. Force Analysis provides a quick and accurate view of what is driving and what is resisting one's efforts.

Card Sort/Affinity Diagram

Originally the name of this tool was the Card Sort; today however, it is most commonly called the Affinity Diagram. The Affinity Diagram is an effective way to stimulate creativity in a group. The Affinity Diagram can also help to provide some structure to a large number of ideas. It works well with sensitive topics since it promotes interaction without criticism. The Affinity Diagram is a synergistic tool that can help break through barriers preventing progress in the past.

Example of the Affinity Diagram

STEP 1. Define the opportunity, problem or situation in very broad terms. Word the definition clearly.

STEP 2. Generate ideas using Brainstorming. Each idea should be at least three words, including a noun and a verb. Write each idea on a separate card. Place only one idea on a card, and write clearly so everyone can read the cards.

STEP 3. Sort the idea cards into groups. Sort the cards at the same time and in silence. Sort the cards by moving a card next to another card to which it is related. Moving a card someone else has placed and putting it in a new place is allowed. Continue sorting in silence.

STEP 4. Generate header cards. Talking and discussion are encouraged during this step. Generate a header that captures the essence of all the ideas of the cards in each group. Each header statement needs at least three words, including a noun and a verb. Headers must be as specific as possible. Write headers and place them above the cards in the category. Mark header cards to distinguish them from the idea cards. The header can be marked with a letter or number.

The Affinity Diagram tool takes the politics out of developing ideas, since everyone has equal power using this tool.

Mind Mapping

Mind Mapping is a creativity technique that integrates the processing of the whole brain. It promotes the visualization of ideas and provides a method to expand creativity by balancing the influence of logical evaluation. Mind Mapping is very effective in helping people break through old paradigms using the intuitive powers of the mind. Since it is both visual and logical, it aids in the generation of creative alternatives. Mind Mapping uses color and images to invoke the right brain and break word-oriented left-brain tendencies. Selected use of words integrates the left brain into the mind mapping process.

One or two individuals usually complete Mind Maps, but they can incorporate the creative ability of many people working together if the formatted workspace is structured appropriately. Groups can work on the same map on a chart pad or board or build new images from the images of others. Mind Mapping is also useful for taking notes or outlining a book.

Example of Mind Mapping

STEP 1. Define the topic.

STEP 2. Draw a central image of the topic. Use color and a symbol to involve the creative right brain.

STEP 3. Record related images around the central image.

STEP 4. Follow the same technique for each of the new images. Let the mind wander just as it did with the central image. As other images come to mind, record them around the image from which they were generated. When the process of generating images is completed, connect each image to its related image. Remember that a single word can be used as an image if no other image will come to mind.

STEP 5. Expand images as long as creativity continues.

STEP 6. Group together ideas that have common themes by drawing a colored line around all the images in a group, marking them with a code, or redrawing the map to cluster common items together.

Expand every image on the Mind Map until ideas are exhausted. As Mind Mapping is employed, drawing meaningful images becomes easier. Using the Mind Map is a method that works best for individuals working on personal issues. It uses a provoked style of creativity and is best applied to planning. The creative power of this tool is best seen when it is applied during medium time frames. It provides a powerful method to unleash the power of the right brain.

Environment

One of the most important and overlooked creative tools involved in a formatted workspace is the environment. The environment includes the physical space, the furnishings of the space, the lighting, the music, and other surroundings. "Aha" creativity is encouraged by changing the environment.

Break out of the comfort zone and make changes to where work life and home life occur. Change associates, listening music and reading material. One's paradigm is defined as his or her approach to situations. In order to become creative, one's paradigm must be broken, and opportunities, problems and situations must be viewed from different perspectives.

Some ways to change one's environment are to take a walk, listen to music, or take a nap. Do the unexpected, such as working in a hallway for a day or two. This tool is one reason that so many deals are consummated at the golf course or in a club. The change in environment enables people to see things differently and be more creative.

A formatted workspace can help a person to be appropriately creative. When determining one's approach to creativity, a formatted workspace should be a prime consideration. Not having the appropriate formatted workspace curtails a person's ability to be creative. Formatted workspaces provide structured ways of achieving creativity using a person's preferred thinking approach.

To maximize creativity, a formatted workspace must:

- Combine different thinking approaches, forcing people to use more of their whole brain.
- Include the physical characteristics of the space in which working and thinking occur.
- Use different methodologies to stimulate creative forces.
- Be fun.
- Stimulate all of the senses.
- Allow both sides of the brain to work together.

A formatted workspace may be a garden or a special room within the home or office complex. A formatted workspace may be a methodology, such as mind mapping or drawing relationships on a flip chart. A formatted workspace can also be something as creative as a guided fantasy, taking a nap, going for a walk, disengaging from the rest of the world through meditation, or merely closing one's eyes and listening to music. Using a formatted workspace evokes insights that you may not achieve in another manner.

7. Stimulating Mechanisms: Focusing Energy

Words either describe the facts or distort them, and facts are always non-verbal.

Mechanisms do exist for using the rich reservoir of words to stimulate creative thinking. Each of these tools can and should be modified to meet the needs of the situation, as not every word evokes the same thought in everyone's minds. Create a personal list of stimulating words, and find the best place where these words can be used. Following are two creative thinking methodologies:

- Five Why's
- Thinking Words
  - Combining Ideas
  - Rearranging/Reversing
  - Exaggerating
  - Adapting
  - Transforming
  - Eliminating
  - Substituting

Five Why's

The five why methodology helps a person systematically discover vital information. Use it for root cause analysis or for asking penetrating questions that require creative solutions.

Example of Five Why Methodology

step 1. Ask why in relation to an opportunity, problem, or situation.

- Why does our product have poor quality, and/or why does the service of our product have poor quality?

step 2. Ask why in relation to the answer to the first why.

- Why was the product not correct, and/or why was the service of the product not correct?

STEP 3. Ask why in relation to the answer to the second why.

- Why was the total process for developing and producing the product and its related systems not capable, and why did the people not have the required knowledge and skills?

STEP 4. Ask why in relation to the answer to the third why.

- Why is the importance of the organization's processes, knowledge and skills not recognized by the business plan and management of the organization?

STEP 5. Ask why in relation to the answer to the fourth why.

- Why does senior leadership not understand and believe?

STEP 6. Continue this process until a point is reached where a creative idea or solution is possible.

- Does the process need to continue?

Using the Five Why method works best for groups working on enterprise issues. It uses a very structured style of creativity, and it is best applied to analyzing tasks. The creative power of this tool is best seen as it is applied within short time frames. This tool is great for getting to the root cause(s) of situations.

Thinking Words (CREATES Method)

Endeavor to know what keeps you within the narrow confines of what is known. Full and correct knowledge of the known takes you to the new and the unknown.

Thinking Words provide a checklist of questions or ideas to investigate. These questions make it possible to apply a structure to one's creativity. Another structure is to use an acronym. A good acronym to use is CREATES. This acronym helps to unleash creative power.

Example of Thinking Words

STEP 1. Isolate the opportunity, problem or situation you want to think about.

- You manufacture traditional wire paper clips, and you want to improve your product.

STEP 2. Ask CREATES questions about each stage of the opportunity, problem or situation and see what new ideas emerge.

Combining Ideas

Much creative thinking involves the combination of ideas. Combining ideas is the process of putting together previously-unrelated ideas, goods or services to create something new. To combine ideas, ask questions such as:

- What ideas can be combined?
- Can purposes be combined?
- Can the combination be an assortment?
- Can the combination be a blend or an ensemble?

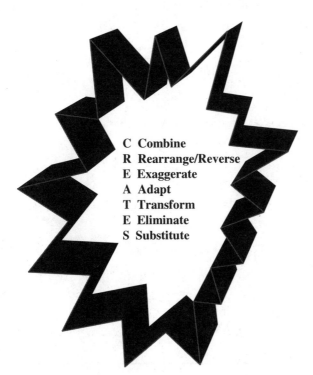

C Combine
R Rearrange/Reverse
E Exaggerate
A Adapt
T Transform
E Eliminate
S Substitute

FIGURE 3.4 Thinking Words

- Can units be combined?
- What other articles can be merged with this idea?
- How can I package a combination?
- What can I combine to multiply the possible uses?
- What materials can I combine?

Rearrange/Reverse

Rearranging what is known in order to find out what is not known is often a part of creativity. Rearrangement usually offers countless alternatives for ideas, goods and services. Reversing one's perspective on ideas, goods or services opens up thinking capability. Look at opposites, and ideas will be seen that are normally missed. To help with rearranging or reversing, ask questions such as:

- What other arrangement might be better?
- Can the components be interchanged?
- Is another pattern or layout possible?
- Can the order, sequence or pace be changed?
- Is the schedule fixed or can it be changed?
- What are the negatives?
- Should it be turned around?
- What if it was pointing down instead of up?
- What if it was backwards?
- Can roles be reversed?
- What is expected?

Exaggerate

Many individuals believe that bigger is better. People often perceive objects they value highly as being larger than objects they value less. Search for ways to exaggerate, magnify, add to, or multiply ideas, products or services. To exaggerate, ask questions such as:

- What can be exaggerated, made larger, magnified, extended, or overstated?

- What can be added?
- What can be made stronger, higher, or longer?
- Can I add extra features?
- Can I do this more often?
- Can it take more time?
- What can add extra value?
- What can be duplicated?
- How can it be carried to a dramatic extreme?

Adapt

One of the paradoxes of creativity is that in order to think originally, one must first be familiar with the ideas of others. Adaptation involves using others' ideas and changing them to satisfy needs. To become an expert at adaptation, ask questions such as:

- What else is like this?
- What other ideas does this suggest?
- Does the past offer a parallel?
- What can be copied?
- Who can be emulated?
- What idea can be incorporated?
- What other process can be adapted?
- In what different contexts can the idea work?
- What ideas outside my field can I incorporate?
- What other uses does this have?
- What other uses does this have if it is modified?
- What else can be made from this?
- Are there other markets for it?
- Can it be extended?

Transform

Just about any aspect of anything can be transformed. Finding those things that can be transformed, and making the transformations, is dependent on each individual. To transform ideas, ask questions like:

- How can this be altered for the better?
- What can be transformed or modified?
- Can a new twist be added?
- How can meaning, color, motion, sound, odor, form, or the shape be changed?
- What new name can be used?
- What other changes can be made (in the plans, process or marketing)?
- What other form can this take?
- How else can this be packaged?

Eliminate

Creative ideas sometimes come from repeated trimming or elimination of portions of the idea. By eliminating portions, the idea can gradually narrow to the part or function that is really necessary. The idea really may be appropriate for another use. Find things to reduce, eliminate, streamline, omit, and miniaturize by asking questions such as:

- What if this was smaller?
- What should I omit?
- Should I divide it?
- What if I split it up or separate it into different parts?
- What happens if I understate it?
- Can it be streamlined, miniaturized, condensed, or compacted?
- Can something be subtracted or deleted?
- Can the rules be eliminated?
- Can the paradigm be changed?
- What is not necessary?

Substitute

Substituting objects, places, procedures, people, ideas, and even emotions is possible and often desirable. Substitution is a trial-and-error method of replacing one thing with another until the right idea is discovered. To find ideas using substitution, ask questions such as:

- What can be substituted?
- Who else can do it?
- What else can be used?
- Can the rules be changed?
- Do other materials exist that will work?
- Is another process or procedure better?
- Can this be accomplished somewhere else?
- Can a different approach be followed to produce the same or a better result?
- What else can be used instead of this?
- What other component part can be used?

Using Thinking Words works best for groups working on enterprise issues. Thinking Words uses a non-linear style of creativity and is best applied to analyzing tasks. The creative power of this tool is best seen when it is applied to short time frames. This tool is ideal for analyzing situations and creating alternatives to something that already exists.

8. UTILIZING EXPERIENCES

Creativity is not a spectator sport. One cannot learn to be creative by watching others be creative; one cannot create by watching others create; one cannot create in a neutral and passive way. Creative thinking requires birthing new ideas, inventing that which did not exist before, and involving oneself in the process.

To break the status quo and move on to something new, physically going someplace and getting out of the workspace is often required. Take a tour and go see something different. See another organization or person. Observe, feel, taste and touch some new ideas in action that will stimulate bigger and better things. Ideas alone will not change the world. Ideas and action are both necessary to change the world.

Reality Matrix

The Reality Matrix provides a check on a process or idea to ensure that it is completely analyzed. The Reality Matrix provides a list of what, who, and when things must be done. This list enables creativity to be unleashed on areas that are not addressed in the current process or plan.

Example of the Reality Matrix

STEP 1. Define the steps to the process.

STEP 2. Define who is responsible for each step.

STEP 3. Define when each step will be done.

STEP 4. Analyze missing steps and completion times.

Using the Reality Matrix works best for groups working on enterprise issues. It uses the non-linear style of creativity and is best applied to analyzing tasks. The creative power of this tool is best seen when it is applied to short time frames. This tool is ideal for finding holes in plans or processes.

Presentation

One's success depends on involving others. To gain support, a person needs to share what has been done and how creative solutions have been birthed. Following the simple guideline of telling them what they are going to be told, telling them, and then telling them what they were told provides very powerful results.

Thinking in this fashion unleashes one's creative potential. Presentations are usually not formal, and people can benefit from seeing non-traditional presentations. Be certain to practice the presentation until the content is both understood and familiar. Be willing to use song, dance, gestures, computers, and any other media that will give the presentation dynamic impact.

Quite often the problem or situation is comprised of a series of impacting and defining forces. These forces can be diagrammed pictorially using tools such as Force Analysis, a Cause and Effect Diagram, or a Radar Chart. Picturing forces need not be limited to a rendering of the problem or situation. Taking a physical tour of where the problem or situation exists is often a good option. Walk through the plant or the factory, and visit competitors, vendors, and customers to get new insights by visually seeing what the relationships are. A picture is truly worth a thousand words. A visit or tour is worth a thousand pictures. Creativity tools yield results.

Be like the chick that pecks at the shell. Speculating about life outside the shell would have been of little use to it, but pecking at the shell breaks the shell from within and liberates it—Nisarga Datta

TAKE AWAY

1. We need tools because our senses of touch, taste, smell, hearing and seeing have a limited range within which they function.
2. The only tool in which we seem to excel is "minding"—our ability to comprehend, memorize, and think.
3. Any sensory information lying outside the ranges of our senses needs to be converted and brought into our range for analysis.
4. Being able to recognize boundaries is an important skill for successful management.
5. Education, a structured work life, and the need for conformity have dulled creative pursuits. Many of us have become like Felix (of "The Odd Couple").
6. Creativity can be structured, non-linear, provoked, or Aha. Each one requires different types of inputs and produces a varying extent of creativity.
7. "Aha" creativity can be described as having no steps or patterns, focused on big issues, having a defining moment, using simple methods, and as individually intense.
8. Mind expanders were designed to help people unleash creative powers.
9. Exaggerating Objectives provides different perspectives to enable one to creatively approach an opportunity.
10. Suspending rules of behavior to gain insight and consider new possibilities is an important tool.

11. Brainstorming may help in generating a large number of ideas in a short time.
12. Mind Mapping promotes the visualization of ideas and provides a method to expand creativity by balancing the influence of logical evaluation.
13. The environment is the most important and overlooked creative tool involved in a formatted workspace.
14. *'CREATES' implies:*
    - Combine
    - Rearrange/Reverse
    - Exaggerate
    - Adaptation
    - Transform.
    - Eliminate
    - Substitute

INNOVATION IN THE INFORMATION AGE

Rajeev Jain

Why are traditional innovation leaders like Ford, Kodak and General Motors struggling today? In the last couple of decades (i.e., the information age), business environment has changed dramatically. Collaboration and technology have significantly increased the pace of competition from local and global firms that can creatively deploy information to provide greater value to customers. The information age has been the catalyst in redefining many traditionally successful business models. As a result, innovation is even more critical and explains many of the differences between leading firms, industries and economies at local, regional, national and global levels.

In many minds, "innovation" sparks images of a long-bearded person with scuffled hair, running across the street screaming, "Eureka! Eureka!! Eureka!!!" In the days of scarce information, innovation was the primary domain of the R&D department and research scientists. Long periods of time were needed to gather information, conduct research, and develop new and improved products and services. Once they were developed, the knowledge leading to the new product or service was not easily available to competitors, and this proprietary knowledge provided the innovating organizations with a competitive advantage for long periods of time.

87

That path to innovation has all changed in the last decade or two. With the emergence of advanced technologies, improved communications, the ease of information-sharing through the Internet, the drive toward globalization, and diminishing trade barriers across countries, the world is a much smaller place where information is becoming ubiquitous. An abundance of information is available to anyone anywhere at any time. Products and services that traditionally take years to replicate are now available in similar form and function within months.

Automobile manufacturers that used to introduce a new model every five years now introduce two or three new models each year. From the time of a single configuration of a micro-computer becoming available to all users (e.g., the IBM XT or the IBM AT in the 1980s), a PC (personal computer) can now be custom-configured by Dell and delivered at almost no extra cost or increase in delivery time. As technology, designs and processes are shared with other parts of the world in search of lower-cost solutions, corporations in those parts of the world are now equally capable of producing the same products and services with much less lag time.

So, has innovation lost its edge? Does it no longer provide the competitive advantage it used to provide? Not at all; in fact, innovation has become even more important. Innovation is still the primary differentiator among the leaders. What has changed is the need for speedier innovation and the generation of these new ideas in greater volume. Today innovation is even more critical and explains many of the differences between leading firms, industries and economies at local, regional, national and global levels.

GOVERNMENT'S PARTICIPATION

Many governmental institutions are realizing the increasing importance of innovation today and proactively trying to create the right environment. The link between innovation and eco-nomic prosperity is unquestionably recognized by both busi-ness and civic leaders. Corporations are increasingly dependent on institutions in their direct environment for their innovativeness and thus competitiveness. The civic leadership can provide the right impetus that can lead to leadership in innovation. Following are examples of such efforts around the globe.

UNITED STATES

In the US, federal statistics estimate that up to 30% of the nation's economic growth over the next 10 years will be derived from innovation. For any region with an economic base of $100MM, and assuming a moderate growth rate of 2.5%, this translates to $7.5 billion for that local economy! For example, northeast Ohio has been a very active region in promoting innovation. The formations of the Greater Cleveland Partnership, JumpStart, Team NEO, and the Fund for our Economic Future are all examples of that community's effort to develop new approaches to improve a region's economic potential. Additionally, the common agenda for all these organizations and programs is to create the infrastructure and relationships to allow northeast Ohio to be more innovative.

GERMANY

Last year, while speaking at the opening ceremony of CeBIT, the world's largest technology fair held in the city of Hanover (Germany), Chancellor Gerhard Schroeder said that Germany needs to develop new sources of growth if it is to achieve sustainable economic growth rates and higher levels of employment. "We are therefore complementing the structural reforms with an innovations strategy," he said. "We are pulling out all the stops to get the innovation system in Germany into gear. That's why I have announced the 'Partner for Innovation' initiative."

UNITED KINGDOM

The Department for the Environment, Food and Rural Affairs (DEFRA) has published its Science and Innovation Strategy that comprehensively sets out the science activities the department proposes to carry out. DEFRA's science minister, Lord

Whitty, said, "We recognize that the science we will need in five- or ten-years' time is likely to reflect emerging priorities and new scientific developments, as well as the goals which we are setting in this strategy. That is why I endorse the forward-looking approach that is proposed here, to examine the likely nature of our scientific requirements over the next ten years in order to focus and refresh this Science and Innovation Strategy and to scan the horizon for new opportunities and challenges." This strategy is considered a key to sustainable development.

CANADA

Federal Industry Minister, Allan Rock, laid out the government's plan for a 10-year innovation strategy aimed at securing Canada's place on the leading edge of technological information, training, and research-and-development. The government leaders believe it is imperative to create a business environment that will attract foreign investment in order to promote new expansion in the knowledge-based economy. This strategy, coupled with the government's allocation of $1.1 billion over three years to support skills development, learning and research, will better enable Canada to attract high-quality foreign investment.

AUSTRALIA

The Australian government committed almost $3 billion to improving the nation's ability to create wealth from scientific achievement. The government leaders said they will target biotechnology and information technology industries. However, the majority of funding will go to education and training. For example, the Australian government unveiled a $1 billion postgraduate loan scheme and thousands more university places for first-year students as part of its innovation strategy.

Just as the leading corporations need to be innovative in their industries, the leading nations must provide the environments that encourage innovation in order for them to remain leaders in the global economy.

The role of innovation in any economy typically addresses four key areas:

1. *Knowledge performance*—Creates knowledge and brings the ideas to market more quickly, as well as increases investment by all sectors in research and development.
2. *Skills*—Ensure that the economy has enough highly-qualified people with the requisite skills for a knowledge-based economy.
3. *The innovation environment*—Modernizes business and regulatory policies to support and recognize innovation excellence while protecting the quality of life.
4. *Strengthen communities*—Support innovation at the local level so our communities continue to be magnets for investment and opportunity.

Some more examples of such an effort, in addition to those shared earlier, include:

- In Australia, the government's innovation plan will include:
  — A premium research and development tax concession rate of 175% for businesses spending money on R&D above a base level determined by recent performance.
  — Doubling of the budget for the Australian Research Council at a cost of $660 million over five years.
  — A boost for the Co-operative Research Centre program by $150 million over five years.
- In Canada, the government in the province of British Columbia made a multi-ministry investment commitment. The government is committed to making British Columbia a world leader in innovation and scientific advancement. British Columbia has an enviable record of technology transfer and commercialization. This Canadian province provided the firm Genome BC with $27.5 million to further its effort in genomics research. New discoveries emerging from the research will lead to spin-off companies, jobs and tools to improve healthcare.

BENEFITS OF INNOVATION

If governments are so keenly stepping up to the challenge, the benefits of an innovative economy must be very visible.

— The OECD (Organization for Economic Cooperation and Development), through its analysis of economic trends, indicates that highly-innovative countries tend to enjoy higher standards of living. Innovation was an important contributor to economic growth in the 1990s.

— The Conference Board of Canada indicates that highly-innovative firms have higher sales, are more profitable, and create more jobs than weak innovators. Innovation is seen as being critical to achieving higher productivity and to attracting more direct investment from abroad by the Canadian government.

— The President of ASME (American Society of Mechanical Engineers) has stated, "Innovation will be the fuel that further creates jobs and strengthens US competitiveness in the global economy."

Such strong sentiments are further supported by formal research conducted by various independent professional organizations.

— According to the strategic consulting firm, Arthur D. Little, top-tier innovating companies created a 12% higher shareholder return (CAGR) over a 10-year period than less innovative companies.

— A July 2001 survey by PricewaterhouseCoopers shows that top executives in companies embracing innovation indicate that the greatest benefits resulting from innovation were found in:

  • new product and service development (83%);
  • revenue growth (80%);
  • higher earnings or profit margins (77%), and
  • increased efficiency of their organizations (72%).

During an economic slowdown, measures to increase a country's competitiveness, generate value and improve safety become even more critical. At such a time, innovation (i.e., creating and adapting to change as well as creating and implementing new ideas to improve productivity, generate more income, and improve the natural environment, health and safety) becomes more important than ever before.

KEYS TO SUCCESSFUL INNOVATION

Achievement of such positive results requires an effective innovation management process.

Pharmaceutical companies are highly dependent on innovation. More than 200 companies from around the world participated in a survey regarding drug discovery, development and production practices. The results of the study were as follows:

— Companies belonging to the top 20% based on product introductions and sales were arbitrarily defined as "most innovative" or "successful."

— Companies in that group generated 47% of their turnover in the past three years from new product introductions.

— The remaining 80% were defined as "less successful" and had, on average, only 36% of their sales based on recently-introduced products or services.

Further analysis of the study data indicates that five factors strongly correlate with success in innovation. The figures in parentheses indicate how much the average scores of the "successful" companies *exceed* those of the "less successful" companies.

• The definite emphasis on professional management of processes (81%)
• The implementation of knowledge management (75%)
• The proper use of project management techniques (34%)

- The clear definition of a strategy for innovation (33%)
- The creation of a culture of innovation (28%).

The two key differentiators above are both related to execution—management of processes and knowledge management.

Daniel Vasella, CEO of Novartis, believes people do a better job when they believe in what they do and in how the company behaves—when they see that their work does more than enrich shareholders.

THE TRADITIONAL INNOVATION PROCESS

While the governments need to play their role by providing the right policies, resources and incentives, the actual innovation really happens at the institutional or corporate level. The traditional innovation process required research and development to be conducted largely in isolation. Firms would normally hold information in a high level of secrecy and not share it beyond the research team. Even today, these traditional "linear innovation" processes can be very well-suited to growing the existing businesses of established companies.

Today, however, when the competitive environment is changing so rapidly, the need for a new approach is becoming a necessity. Recently, a worldwide study on innovation was conducted by the international management consulting group Droege & Comp. AG and the National Association of German Industries (BDI). The study concludes that companies seldom recognize weaknesses in the area of innovation. Nearly half of the companies having difficulty competing through product innovation still consider themselves to be innovative. The problem obviously does not reside in the lack of innovative ideas or concepts, but rather in the inability of the companies to implement innovation.

TYPES OF INNOVATION

A number of attempts to research the innovation process—the type of innovation, the environment that promotes it, the

Processes

Companies must focus on the various R&D processes used to develop the desired products. Focusing on the essential processes ensures that companies can properly manage them to reduce lead times.

Methods

Management should halt work on R&D projects as soon as it recognizes that the projects are not meeting the company's expectations. Cessation of substandard activities permits redirection of resources to more promising projects in the pipeline. Companies often experience "go–no go" issues as painful because of a lack of clear criteria and information supporting consequent decision-making. Projects that survive are more or less undetected because they are not clearly and officially cancelled. Additionally, projects that are cancelled late, and therefore are at high financial and psychological costs (also known as submarines), create many common R&D pitfalls.

Tools

Successful companies learn that knowledge is the key to success. They invest in internal knowledge-sharing and learn from their project experience—particularly their mistakes. They modify their structure and system to avoid repeating those mistakes. Less successful companies do not master the tools necessary to practice professional knowledge management.

Time

Many examples support how creativity is spurred under time pressures—and many people believe that people work best when they are under the gun. Research suggests otherwise, however. When creativity is under the gun, it gets killed (although a lot more may get accomplished).

Each factor identified in the study has multiple aspects which occur differently in different companies. In addition, the research also commented on the challenges that result from lack of resolution. A closer look at each factor provides further understanding of how those factors affect innovation.

UNUSUAL BARRIERS

The research above confirms and describes the more traditional barriers to innovation. Comments by some well-known innovators provide new insight.

— Thomas Fogarty, the inventor of the first therapeutically-used balloon catheter, practices cardiovascular surgery at the Stanford University Medical Center. Fogarty says, "One of the hardest things about innovation is getting people to accept that the way they work just might not be the best."

— Another barrier to innovation is the fear of hiring smart people—people smarter than yourself—because you may feel threatened by them. Mike Lazardis, founder of Research in Motion, a maker of wireless solutions including the Blackberry, lives by that philosophy. He tells people not to worry about their job and then finds people who can do it better than them.

Many product firms think of innovation in terms of product only; hence product innovation is where most of their R&D expenditure is incurred. Most companies ignore the opportunity for innovation in areas like customer service, business models, networking, and the supply chain. Dell Computers became the leader in its industry (even though it markets a product that is highly commoditized) through innovation in inventory management and its supply chain.

RECENT INNOVATION TRENDS

Firms have always depended on innovation processes for their competitiveness. In order to stimulate innovation processes, firms have to exchange information and turn this information into knowledge—in other words, they themselves have to learn.

The information and knowledge that are needed for innovations can be collected both inside and outside the firm.

Due to intensive competition and shorter product life-cycles, all firms are increasingly dependent on external information and knowledge sources to provide direction for their programs. As a result, a firm's innovation processes now more frequently take place in interaction with other organizations, be it with other business partners (such as customers, suppliers or competitors) or with public research establishments (such as universities or innovation support agencies). Innovation processes seldom take place anymore in isolation. The institutional contexts for this "learning by interacting" differ not only nationally, but also regionally and locally.

The advent of the Internet and digital media has increased information-sharing so broadly that isolated access to information is now rare. In addition, the level of specialization in every field has increased—making it difficult for a single individual to have the depth of knowledge in all related aspects of a product or process. Additionally, investments for pure experimentation required in new technologies can be very high. Hence collaborations and joint ventures are more commonplace than in the past. Such interaction allows innovation to emerge in a wide variety of forms and be elevated to a much higher level than previously conceivable.

Take the example of the innovative trend in the European telecom industry. According to Forrester analyst Lars Godell,

> European telcos don't appreciate that the root cause of their malaise is the vertically integrated model of the past 100 years. With an industry in crisis, traditional merger and acquisition and restructuring within the outdated vertical organizational model can't solve the telecom industry's persistent problems with innovation and missed opportunities. Instead, layering will fix European telcos' fundamental problems as they abandon vertical integration and move to a horizontal structure that will foster innovation, delight customers, tap new sources of demand and speed up organizational processes.

Today many firms have customer councils, supplier forums and dealer conferences as part of the normal business process. The thought behind the investment in these information-sharing media is to learn more rapidly, leverage knowledge from the business network, learn about competitors in the industry, and track changes and trends that will impact the firm. As we move toward a more global world, an increasing number of established companies are creating various networked innovation activities to grow their existing businesses and be alerted early on to opportunities in new, immature markets.

ANALYSIS OF INNOVATION

The last decade's transition to an Information Age has changed how business is done. Corporations are exploiting the wave of technological change and have skillfully displaced companies that traditionally dominated the automobile, computer, machine tool, semiconductor, steel and xerography industries.

Successful companies find ways to innovate in today's information-intensive environment. Knowledge and resources are critical to their success. The speed at which information travels and changes has also increased exponentially. The interchange of information across multiple disciplines has spurned more creative solutions in almost every sphere. New fields of study have spawned bio-medical engineering, genetic engineering, biotechnology and nanotechnology that combine knowledge from previously-unrelated fields.

The investment in innovation is increasing. R&D spending in the United States increased at the rate of 6.1% on average between 1995 and 1999, and then jumped 7.9% in 2000 for a total investment of $264 billion. Furthermore, 35,000 new consumer products were introduced in 2001, a 133% increase from ten years earlier when 15,000 new consumer products were introduced. This trend continues to grow, creating more opportunities for competition and alliance.

Table 4.1 lists some more examples of industries that are changing due to the information age as well as disruptive technologies that are being introduced:

Software is and will be at the core of most disruptive innovations during the next several decades. The World Wide Web has already stirred up imaginative possibilities for a large number of new markets, products and services—all of which are

TABLE 4.1. Industries Experiencing Change Due to the Information Age

INDUSTRY	ESTABLISHED TECHNOLOGY	ADVANCED TECHNOLOGY IN INFORMATION AGE
Photography	Photographic film (silver halide)	Digital photography
Telecommunications	Wired telephones	Mobile telephones
Retailing	Traditional retail outlets	Online retailing
Computing	Computers	Handheld personal digital assistants
Printing	Offset printing	Digital printing
Financial Services	Full service stock brokers	Online automated stock brokerage

based on software. These new markets, products and services will grow exponentially as more and more minds interconnect to utilize them. But startling as these prospects are, they provide only glimpses of the many opportunities that software innovation presents. When combined with software's capacity to learn on its own, create new solutions, deal with inordinate complexities, shorten cycle times, lower costs, diminish risks, and uniquely enhance customer value, effective software management has now become the key to effective innovation for any company or institution. Innovators who recognize this fact will have a genuine competitive advantage. Managers who ignore this caveat do so at their companies' peril.

Technology is one of the elements that can create disruptive change in an industry. People, processes and organization structures are other key elements that can change the industry through innovative insights. Such examples of value innovation arise in diverse industries. Table 4.2 provides names of corporations that have offered leadership in their respective industries through non-technology-based innovation.

These corporations have competed in the most competitive of the industries and created value for their investors. They were not part of the most attractive industries, and they did not make huge commitments to the latest technology. Instead, these

TABLE 4.2. Leaders in Non-Technology-Based Innovation

CORPORATION	BUSINESS
Barnes & Noble	Book Retailing
Charles Schwab & Co.	Investment And Brokerage Account Management
CNN	News Broadcasting
Compaq	Computers
FedEx	Courier/Package Delivery
Home Depot	Home Improvement Retail
IKEA	Home Products Retail
Kinepolis	Cinema
SAP	Business Application Software
Southwest Airlines	Short-Haul Air Travel
Wal Mart	Discount Retailing

high-performing companies are united in their pursuit of innovation outside of a conventional context. In other words, they do not pursue innovation as technology; rather they pursue innovation as value. The above companies achieved quantum leaps in some aspect of value; many have nothing to do with new technology. These companies are value innovators, because they use good and timely information as a key factor in discovering new ways to conduct their business leading to their success.

CHALLENGES IN THE INFORMATION AGE

Collaboration and technology are definitely providing new direction to the way innovation is conducted in the information age. These changes create new challenges for the innovative corporations. Some of the key challenges include:

INCREASED PACE OF COMPETITION

Established companies are being forced into adopting fundamental changes in their approach as well as developing more innovative and efficient business models that incorporate new technologies. These companies are challenged in their search for new ways to innovate with speed. Consequences of not recognizing this change can be catastrophic. Take the example of Encyclopedia Britannica, a leader in its field. It dismissed the competition from Microsoft Encarta until it was too late and lost a very significant portion of the market. While a set of *Encyclopedia Britannica* sold for over a thousand dollars, the digital Microsoft Encarta on a compact disc sold for less than $50 and satisfied a very large component of the consumer need.

MORE EXPERIMENTATION

As the speed of innovation has increased, more options are available, and thus ignoring competing alternatives is more difficult. Experimenting with multiple options is becoming a critical element of the process of change. However, it requires continuous and larger investment.

LOWER COST GLOBAL LOCATIONS

Improved global communication infrastructure, greater capability of lower-cost locations around the world, and reduction in global trade barriers are driving the needs for new skills in global project management and workforce diversity understanding. Such changes are creating job security concerns for Americans.

GREATER NEED TO DIFFERENTIATE

Technology is commoditizing products in many ways. The following comment by a mass merchant buyer is revealing how technology is creating an economic environment close to the *perfect competition* in a market: "It continues to be a challenging retail environment and there doesn't seem to be an end in sight. Our business is being driven by extremely aggressive pricing." The capability of consumers to be able to compare prices online makes differentiation very difficult from one store to another.

CONTINUOUS NEED TO FIND WAYS TO ADD VALUE

Visibility of products, services and prices online makes it very easy for competition to emulate the business model. Thus companies must continuously find innovative ways to add value to the customer. Because a competitor can copy the business model in a relatively short period of time (in six to twelve months), what may be an unique value proposition today becomes the qualifying standard the customer demands tomorrow, forcing the need to continuously innovate.

Examples abound of innovation in the information age using collaboration and technology. The credit card industry faced significant challenges with fraudulent use of credit cards. This problem created high costs to the credit card issuers like American Express, Visa, Master Card and Discover. A company developed a process for performing predictive analysis based on the spending pattern of individuals using credit cards. Based on historical patterns of card usage, any unusual activity was flagged. This innovation has helped limit the cost from fraudulent usage. Today, the technology used to perform this analysis, Falcon (Fraud Activity Loss CONtainment), analyzes 85% of the debit and credit card transactions in the US (and 65% globally).

Hewitt Associates, a leading firm in the human resource field, uses a similar technology to undertake predictive analysis regarding employees. For most corporations today, people are their biggest investment. Being able to maximize the return on their investment in human capital (as well as being able to differentiate them from their competitors) is of great benefit to Hewitt's clients. Based on a deep understanding of the employee characteristics (data variables), they can help their clients understand and answer questions such as:

— Which employees are potential future leaders for the firms, and what is the best way to retain them?
— Which employees are likely to leave the firm?
— What impact do different HR programs have on various groups of employees, and what is the optimum way to allocate funds to various programs?

This combination of collaboration and technology has helped create a competitive advantage for Hewitt (and its clients) through innovation in the information age.

While the information age has created new challenges, it has also provided unique opportunities to differentiate and add value that did not exist before.

RESPONSE TO CHALLENGES

Companies that do not take charge of their innovation processes usually do not profit from innovation; they expend money on uncoordinated projects that sound like innovation but do not result in new revenue, profits or market share.

Especially when resources are limited and competition is intense, companies can benefit from assessing which structures, incentives and investments in innovation are required and how to extract value from them. After assessment, their target should be to improve their ROII, or Return on Innovation Investment. The ROII, however, is not an easy measure to track and report accurately.

Corporations are responding to new challenges presented by the information age in different ways.

RESTATEMENT OF CORPORATE VISION

In the book, *Built to Last—Successful Habits of Visionary Companies* by Jim Collins and Jerry Porras, the book lists companies, stating their purpose as something more holistic. For example, Johnson & Johnson's vision is to *alleviate pain and suffering*; 3M's vision is to *solve unsolved problems with innovation*; Mary Kay's vision is to give *unlimited opportunity to women*; Merck's vision is to *preserve and improve human life*. Hewitt Associates recently restated its vision as *making the world a better place to work*. In these companies, the corporate purpose is then shared and owned by every employee in the organization.

Increased collaboration and global R&D networks

In a recent study of the pharmaceutical industry (which is one of the highest investors in research), the "Best in Class" pharmaceutical companies mastered three process drivers that characterize all R&D innovation: development time, hit rate, and adherence to schedule. Hit rate is defined as the percentage of process runs needed to achieve the desired outcome. Innovation processes are most successful when the development time is short; the hit rate is high, and there is adherence to the schedule. The study showed that 91% of the more successful companies have shorter average development times; 87% of the more successful organizations have a higher hit rate than less successful companies, and 94% of the innovators deliver their innovation on schedule. Interestingly, companies that organize their innovation process internationally report a higher success rate (91%) than other study participants (52%). The internationalization of R&D activities offers intercultural incentives and opens the door to many synergies. Companies can reduce product development time significantly when innovative activities take place around the world and, thus, around the clock. Furthermore, an international R&D network taps centers of excellence that are not necessarily available in a single country. Diverse cultural backgrounds make lateral thinking much easier and provide the "sparks" needed to ignite powerful innovative thinking. Organizations involving their suppliers in the development process tend to be more successful than others (59% versus 48%), although that criterion is less significant because most pharmaceutical companies are vertically integrated upstream.

Mergers to reduce cost

Tough market conditions have triggered global industry consolidation. Pfizer's US$60 billion merger with Pharmacia to create the world's biggest drug company is a prime example. Pfizer faces a challenge. The company expects to spend US$5.3 billion in research for new treatments in the year 2006 alone. At the same time, governments worldwide are pressuring Pfizer to drop prices. Pfizer's last blockbuster, Viagra, was produced four years ago, and it cannot afford too many research flops. A merger appeared to be the only option. Pundits predict that more deals will come in this industry.

While firms are responding with these strategies, the road ahead is not easy. Historical leaders find it difficult to create new relationships or redefine their strategy in light of the changing business environment.

Unilever distrusted collaborative ventures with other firms, especially when those firms had links to its competitors. This distrust meant little effort was made to tap into outside sources of technology available from larger suppliers. The culture of self-reliance, coupled with the internal divisions and communications problems, contributed to Unilever's continuing problems with innovation. Certainly Unilever was not alone in facing these difficulties. Due to its size, diverse markets and geographical reach, the challenges it faced were more acute than many of its competitors. However, the company was reflexive about its shortcomings in innovation, and many within the company both recognized, and were critical of, the internal factors that contributed to the innovation problem.

According to the strategy consulting firm Bain & Company Inc., 42% of the companies that started networked innovation activities last year will halt them because of major write-offs and a lack of the required skills. Jumping into relationships without a clear vision, proper planning and requisite resources is a recipe for disaster. Even if you are late to recognize the change, do not take the reverse approach of "shoot, ready, aim!"

A survey of some 200 manufacturers also identified several major obstacles facing U.S. manufacturers, including a lack of shared vision within companies, misaligned performance and incentive measures, poorly-educated workers, environmental regulations and lack of trust by suppliers. More than half of these manufacturers said they begin product manufacturing feasibility tests only after most of the design work is done.

Improvements to American goods grow, but cost efficiencies and new-product gains do not, says the Boston University School of Management Manufacturing Roundtable. Its latest survey finds that the quality has improved at an annual rate of 7%, up from under 6%. Cost and new-product introduction gains, however, remain stagnant at 2% to 3%, trailing Japan. Continued focus on broader innovation is critical. The innovation process needs to be imbedded into the organizations culture now more than ever before.

TAKE AWAY

1. The link between innovation and economic prosperity is unquestionably recognized by both business and civic leaders.
2. The innovation in any economy addresses knowledge performance, skills, the innovation environment, and strengthening communities
3. Highly innovative countries tend to enjoy higher standards of living; thus innovation was an important contributor to economic growth in the 1990s.
4. Highly innovative firms have higher sales, are more profitable, and create more jobs than weak innovators.
5. Innovation will be the fuel that further creates jobs and strengthens the global economy.
6. Over a 10-year period, top-tier innovating companies create a much higher shareholder return (CAGR) than less innovative companies.
7. The greatest benefits resulting from innovation include new products and services, revenue growth, higher earnings, and increased efficiency.
8. The traditional innovation process required R&D to be conducted in isolation and secrecy.
9. Nearly half of the companies having difficulty competing through product innovation still consider themselves to be innovative.
10. There are three extents of innovation—incremental, architectural and radical. Examples include a better ceiling fan (incremental), the portable fan (architectural), and air conditioning (radical).

11. Most companies ignore the opportunity for innovation in areas like customer service, business models, networking, and the supply chain.
12. Due to intensive competition and shorter product life-cycles, all firms are increasingly dependent on external information and knowledge sources to provide direction to their programs.
13. Today, many firms have customer councils, supplier forums, and dealer conferences as part of the normal business process. These firms use these resources to learn more rapidly, leverage knowledge from the business network, investigate competitors in the industry, and track changes and trends that will impact them.
14. Corporations like Home Depot, IKEA, Southwest Airlines and Wal-Mart are providing leadership in their respective industries through non-technology-based innovation by competing in the most competitive of the industries and creating value for their investors.
15. Companies that do not take charge of their innovation processes usually burn up money on uncoordinated projects that sound like innovation but do not result in new revenue, profits or market share.

NEED FOR INNOVATION ON DEMAND ·

Abhai Johri

Once innovation is embedded into an organization's culture, a disciplined and structured thought process becomes part of the daily routine. The employees constantly generate new and creative ideas that contribute to an organization's success and its competitive advantage. For such organizations, innovation is not a special effort but a way of life—innovation on demand. However, building such a culture requires organizational commitment to create the right environment as well as a conscious removal of barriers. This is not a simple task, as it requires a commitment of financial and human capital— but the rewards are worth it.

In today's global, multicultural and competitive world, why are some organizations extremely successful, other organizations struggling to survive, and still others failing to stay in business? Of course, many factors abound in this chaotic and unpredictable world for getting varied degrees of results. For example, the turbulent environment, intensifying competition, demanding and sophisticated customers, shorter-time horizons and a discontinuous future all play a part in the outcome of a company's success.

Several other business factors may also influence the results of business success, such as the availability of resources (financial and intellectual), the quality of products and services

111

offered, uniqueness of the offerings, strength of positioning and marketing, establishment of a brand image, promotion of discriminators, cost management, and simply staying in a complex cultural environment. When such a multi-dimensional set of variables is present, thought must be given as to what should be done differently to be successful. Asking and answering such questions as, "What do we represent?" "What is the company's purpose?" and "How does our unique identity get established in the marketplace?" is important.

After analyzing some of the successful companies that have existed and thrived for a long time, two fundamental themes stand out. First of all, such companies have well-defined *core values* that are never compromised. Secondly, such companies practice well-established *habits* to continuously change, adapt and evolve to the marketplace and its demands.

To continuously stay on track with the fast-paced world and its demands, to remain above the competition, and to continue to offer unique advantages to clients, an organization must possess a carefully-crafted process that can be used frequently (or when needed) to change and improve. This *process of innovation* should be well-integrated within the organization's culture, such that it is frequently used as part of the ongoing activities. The *habits* for disciplined and structured thought processes provoke new ideas and creativity that are in heavy demand in successful and long-lasting corporations. These habits and process are referred to here as *Innovation on Demand*.

Traditionally, innovation was treated as the art of geniuses—those who possess desire, discipline to always think proactively, and an ability to go against accepted norms. In addition, such individuals are risk-takers who are never shy about trying out new things or facing failure. Such people have vision and passion in different fields, such as science, art, politics, economics and philosophy. For example, Gandhi initiated the war of independence using only a non-violent approach (totally unheard of before). With extraordinary vision and confidence, Gandhi won the war without any bloodshed. Albert Einstein had a passion toward unleashing imagination that resulted in the most revolutionary invention—the *Theory of Relativity*. These innovations were self-generated.

Unlike these innovations, the innovations discussed here do not require geniuses. In many cases, lacking a genius in the team is actually more desirable when discussing how to innovate based on defined guidelines within a group of people. Lael Lund stated, "*Innovation* is about making something better. It is about having a vision for an idea, spotting the gaps that are in the way of achieving that vision and coming up with a relevant, practical, and usable product at the end." At its broadest, *innovation* means finding new or better ways to do things, creating new products or solutions, applying new technologies to solve existing problems, or using existing technologies and products to meet new needs.

Innovation on Demand is a process that enables an individual, or a group of individuals, to follow a structured approach to think, discuss and arrive at a set of new ideas that are unique. Such ideas position the organization with unique discriminators to contribute toward its performance. This kind of innovation has a purpose, a strategy for survival and growth, and a defined environment.

INNOVATION AND CREATIVITY

Innovation is the *implementation* of a creative idea that is new and unique. Focusing the creative power of researchers and teaming them with subject matter and implementation experts can bring tremendous success for a company. Contrary to popular opinion, creativity is not just an art, but it is a science as well. While everyone may not be a Nobel Prize winner, everyone possesses creative power that can be accessed using well-defined processes and techniques. The traditional technique of *brainstorming,* and the extremely popular and successful technique of *Six Thinking Hats,* can be used to improve creative abilities.

Creativity is a mental gym, which when exercised brings useful results and strengthens thought processes. Creativity requires much discipline and a willingness to look for alternatives. In order to be successful in its environment, an organization should encourage two types of creative activity for its employees:

1. **Divergent/Lateral Thinking** (mentioned in Chapter 2) utilizes creative forces to identify several new ideas and options. Lateral thinking focuses on generating alternatives. Such thinking is capable of disjunctive shift (where there is no wrong way). Divergent/lateral thinking is also provocative, in that it forces people to think of alternatives based on a set of ground rules.

2. **Convergent/Vertical Thinking** utilizes a logical process to determine the best solution. Vertical Thinking focuses on selecting the best (or the optimal) alternative. It is sequential thinking that must be correct at every step. It is also analytical thinking, in that it forces people to make an optimal choice based on a given set of ground rules.

An organization should include both types of thinking approaches in a process for innovation. Such thinking approaches promote the generation of new ideas for a specific organizational purpose, such as how to position the organization against a new product in the market. Lateral thinking will aid in the development of a list of alternatives and options. Vertical Thinking will help with the selection of the best alternative among all possible alternatives.

A set of guiding principles and goals for those involved in the innovation process should be defined. Finding the best solution that everyone in the team can agree upon is sometimes difficult. In that case, guidelines to find an optimal solution that the majority can agree upon are very helpful. The ex-CEO of General Electric, Jack Welsh, initiated and promoted the *Innovation on Demand* process throughout the company. Welsh used *Innovation on Demand* to periodically solve complex problems using a selected group of people. He sometimes imposed tough rules, such as "a team is locked together until it reaches a resolution for the problem." This approach resulted in tremendous success for GE in launching new products and implementing new business strategies.

Six Thinking Hats (as discussed in Chapter 3) is a simple and extremely powerful tool to structure creative discussions among people. This tool allows the group to think about one thing at a time rather than juggling several balls in the air. Each thinking hat is focused on various aspects of thinking—facts, emotional view, negative view, positive view, creative (different) view, and control view.

CREATIVITY BLOCKERS

Although geniuses are not the only ones able to generate innovative thoughts, fear of being creative is still present in ordinary people's minds. The following are some of the common blocks to creativity that hinder people from generating creative thoughts:

- *Fear of failure* (success rewarded/failure punished)—Here people are reluctant to take risks and innovate because of their fear of failing (thus possibly damaging their reputations or experiencing punishment).
- *Allergy to ambiguity* (uncertainty = chaos)—If the problem definition is not clearly defined, people are sometimes unwilling to try. They are conservative and do not want to take the risks if they do not understand the problem(s) completely.
- *Touchiness* (self-worth, fear of rejection, etc.)—Some people are sensitive to their egos and do not want to expose themselves by participating in a creative process.
- *Conformity*—Some people would like to stick to the action or behavior in correspondence with socially-accepted standards, conventions, rules, or laws. They do not want to think or do anything that is out of the traditional norms. This fear suppresses out-of-the-box thinking and thus creativeness.
- *Resource myopia*—Some people are just too lazy to exercise their creative power; they may also be short-sighted and not able to look beyond a close horizon.
- *Starved sensibilities* (Lack of exposure)—If people have limited exposure to the world, this lack of exposure may limit their ability to think beyond their own knowledge and experience. As Raymond Inmon said, "If you are seeking creative ideas, go out walking: Angels whisper to a man when he goes for a walk." Walking, with its constant inflow of new images,

gives us new thoughts. Thus people should make an effort to explore in order to succeed.

- *Rigidity* (Fixed frames of reference)—People may be so biased toward predetermined notions that they cannot or do not think outside that frame of reference.

UNBLOCKING CREATIVITY

The organization should use different strategies for unblocking these fears that are a deterrent to creativity and continue to create an environment that encourages creativity and innovation. The following are some of these strategies for unblocking such fears:

- *Awareness*—Increase the awareness of the organization's expectation and demand for innovation, and provide adequate training to start thinking more proactively.
- *Analysis*—Introduce the analysis technique, using it on creative techniques that can achieve similar results to the current status quo.
- *Help from credible sources*—The fear of risk-taking and failure can be significantly reduced if people are facilitated by outside experts who specialize, and have extensive experience, in motivating people toward innovation.
- *Inoculation* (a little at a time)—The inoculation strategy is a preemptive advertising tactic in which one party attempts to foresee and neutralize potentially-damaging criticism from another party by being the first to confront troublesome issues. This strategy can be done through a facilitator in order to help people overcome their fears.
- *Rewards*—Introduce rewards for submitting new ideas that can provide benefits to the organization. This strategy has been used by several organizations over the years and has proven to be successful in generating innovative ideas.

In this competitive world, an organization must establish an environment to demand and generate creative activities from its employees. These creative activities provide collective thoughts that can help continually improve the organization's position while maintaining its uniqueness, and increasing its strength of offerings, in the marketplace.

CONVENTIONAL INNOVATION CYCLE

Everything is a product of past innovation. Products are created by ordinary people—not just the genius. They are produced by a conventional cycle of innovation, where the entire task of innovation is divided into several sub-tasks. Each sub-task can then be performed by people with different skills. Thus experts in all areas are not needed.

The conventional innovation cycle enables the generation of innovative ideas by the efforts and inspiration of a genius. The *Innovation on Demand* process creates the environment and processes whereby ordinary individuals are trained in specific areas of the entire innovation cycle and are given responsibilities to focus on a specific set of sub-tasks. In such an environment, division of the tasks and group efforts collectively produce innovative results that achieve the organization's objectives.

In an organization, the structured and disciplined generation of innovative ideas is needed at various levels, including business strategy, solving client problems, and inventing and implementing new and improved products, solutions and services. Typically, an innovation cycle for building a product and business is comprised of the following sub-tasks:

1. DEVELOPING BUSINESS STRATEGIES

One example of *Innovation on Demand* at the business level in an organization is the process for developing and executing business strategy. Most visionary organizations establish their business strategy focused around their short- and long-term goals. Periodically, they have their corporate strategy team meet to review current strategy and brainstorm to generate new ideas based on industry trends, market opportunities and client requirements. At such meetings, the corporate strategy team makes necessary adjustments in the current strategy when required. Sometimes, the team may bring in outside experts who can provide input on the state of the external environment, facilitate generation of new ideas, and converse on the finalization of the new and improved business strategy. An innovative approach in establishing an evolving business strategy is the key for business success and the growth of any organization.

Innovation in the market strategy itself is very important in order to produce product discriminators as well as position the company correctly and differently from other competition. Developing innovative ideas and concepts is essential in order to market products that discriminate them from the competition and communicate them as more attractive to the users. Based on different marketing strategies, two similar products can have totally different results. Therefore, innovation at every step of the process is equally important to achieve success in this highly competitive marketplace.

2. Conducting Research and Development

Most successful organizations make a fair investment in the Research and Development (R&D) activities to ensure they continue to invent new and improved products and solutions to their clients as well as maintain a competitive edge in the marketplace. The R&D activity helps establish a structured environment that demands and encourages creativity and innovation focused around organizational objectives and goals from ordinary people—not just the geniuses. In today's competitive world, where the technological revolution is changing the world so fast, continuing to invent new and improved products that can differentiate an organization from its competition, and also offer better and more cost-effective products to clients, is imperative. Creating an excellent R&D structure establishes an environment where individuals with specific skills are assigned and trained on a sub-set of the tasks and given a specific set of actions to be performed that are focused around a well-defined set of objectives. The collective actions and tasks performed by the group of individuals in the R&D community produce innovations that can satisfy the organization's objectives for ongoing innovation.

The R&D activities require performing market research and other searches to determine what is available in the market in the area where producing a new product or solution is required. Based on the results of such research, determining the gaps between what is available and what is required is possible. Determining these gaps will help in developing a set of new ideas to produce something unique rather than producing a "me too" product. These ideas should be revalidated against the business focus of the organization to ensure they fit within the business scope, as well as to determine whether the organization has the resources and budget to pursue the idea.

3. DEVELOPING PROOF-OF-CONCEPT

Once the idea has been revalidated and approved, the next step is to develop a sample prototype product or solution using a quick method to test the feasibility of the idea. This process, called *Proof-of-Concept*, helps in testing the idea without investing full efforts and time in building the finished product. This process will also help seek early feedback from typical users on the functionality and usability of the product that can help refine the product based on the actual needs and its usability.

Once the prototype has endured the iterations of feedbacks and improvements and is ready for actual product development, further revalidation of the market size against the financial investment for producing, marketing, distributing and selling the product is required to ensure reasonable and required Return on Investment (ROI).

4. OPERATIONAL EXECUTION

Once the above conditions have been fulfilled, the actual process of developing and manufacturing the product starts, along with the parallel activity of business development, to ensure that the product is produced with good quality and specifications and sold and delivered to the consumers. The execution step should also take an innovative approach to manufacturing in order to increase efficiency, reduce cost, and maintain good product quality.

For example, Japan took early initiatives in mastering its manufacturing process for producing consumer products and thus dominated the consumer products market worldwide for a long time. China followed the lead and is now dominating this market by producing good-quality, very-low-cost products worldwide. In this global economy, companies are taking different initiatives in producing low-cost, high-quality products. This includes offshore development and technical support in order to find the best value (cost and quality).

THE INNOVATION S-CURVE

The famous S-curve as shown in Figure 5.1 applies very well for innovation. The tail end of the S-curve shows that there is a slow progress at the start of any innovation cycle when brain storming and other idea generation activities are performed. As time progresses and innovative idea convergence begins to occur, the innovation process proceeds at a dramatic speed, thus quickly causing the steep slope of the S-curve. At the top of the S-curve, innovative idea generation for the given concept has achieved its

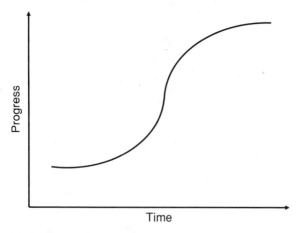

FIGURE 5.1 The S-Curve

zenith, and continuing the innovation process would not produce the same proportional improvement. This zenith is an indication to stop the innovation process for the given idea, as continuing with the process will provide diminishing or slow returns. At this time, applying out-of-the box thinking to determine new products or approaches is appropriate.

For example, IBM has started to focus its growth on *Innovation Services*, where it can continually analyze clients' requirements, and their evolving and changing businesses, based on the market trends. IBM's aggressive initiative for On Demand Innovation Services (ODIS) is a capability that enables businesses to tap into one of the world's most prominent groups of researchers and scientists, who bring with them a set of innovative tools, solutions and expertise in cross-industry business areas to solve complex business issues. Utilizing ODIS ensures that before a company reaches the saturation point on the S-curve, it generates new ideas to move on to new products and solutions required by the marketplace.

IBM's ODIS can bring business transformation insight to an organization that can enable it to implement changes that impact strategic direction as well as operating and competitive models. ODIS researchers and consultants have partnered with clients to find solutions for many tough problems, including analyzing and optimizing value chains, improving collaboration with suppliers, improving inventory management and demand forecasting, and analyzing customer information databases to identify the most profitable customers and improve customer service.

INNOVATION IN TECHNOLOGY

Various types of technologies, from *low-tech* to *high-tech* to *super-tech*, exist. Implementing familiar technology (such as building roads) is low-tech, whereas landing on the moon is an example of super-tech. More innovation is demanded as progression from low-tech to super-tech occurs. If a business is involved in a super-tech project, such involvement will demand more innovation in new ideas for new products and solutions to solve complex problems.

Across the world, innovation is recognized as the single most important ingredient in a successful modern economy, giving high rates of return on investment and driving economic growth, high quality of jobs and high standards of living. Science, technology and innovation are key drivers of future prosperity and quality of life in any country. Several countries around the world are heavily investing in Science, Technology, and Innovation initiatives. Such initiatives are acting as a catalyst for knowledge and wealth creation by enhancing the science and technology base and by facilitating the delivery of beneficial research outcomes. Innovation delivers value in the form of new products and production processes as well as better-quality goods and services. Innovation also benefits the wider community through advances in vital community services, such as healthcare, education, communications and transport.

Several technological areas are continuously evolving and moving extremely fast, especially Information Technology, Electronics, and Biotechnology. Businesses and countries must stay at par with technological innovation to stay competitive in the highly competitive global world. They need to offer products and solutions based on the latest technologies and utilize technologically innovative ideas in manufacturing to produce the best quality products at the lowest cost.

Scientists and innovators are combining various traditional technologies to devise new techniques, products and solutions. For example, nanotechnology was invented as a new technological area that combines the concepts of bioscience, microelectronics, physics and computer science to develop faster super-computers that can help solve complex (NP-Complete) problems in real time with higher probabilistic bounds. For example, higher computing power will help in providing more accurate weather forecasting and finding a cure for chronic diseases (cancer, AIDS, etc.) that was previously impossible to accurately solve in real time.

Worldwide, universities, companies and countries have established centers for business advantage through the innovative application of technology and services. These centers take a broad view of innovation by being the catalyst for stimulating, fueling and enabling new opportunities for today's technology professionals. These centers serve multiple vertical industries, including financial services, healthcare, information technology, interactive media and nanotechnology as well as some of the most forward-thinking companies in this market. By utilizing these centers, technology professionals can grow their networks and expand their reach in this ever-changing market. These types of interactions will increase the knowledge among professionals of the research and trends in the marketplace and thus help in the generation of new innovative thoughts by these professionals.

Innovation in Product Development

In this highly competitive marketplace, continuously improving products in terms of quality, functionality and marketing strategies is essential in order to increase market penetration and continually receive customer acceptance and demand for products. MIT has a well-established Center for Innovation in Product Development (PD), where scientists are performing research on various disciplines of PD, including manufacturing, marketing, design, finance, sales, strategic planning and more. The variety and complexity of these specializations require that PD must be shared by many persons and organizations.

A typical example of product improvement is in photographic cameras. With the revolution in optical, mechanical and digital technologies, cameras have made significant functional, quality and usability improvements in recent years. Cameras have progressed from optical changes, such as zooming, wide-angle capturing of pictures, and infrared focus and lighting, to live videos to digital technology implementation, adding exceptional and unlimited features. With a very competitive yet huge marketplace for cameras, each manufacturer is investing lots of money in innovating new features, functions and quality to ensure its products thrive. The manufacturers are applying the innovations in electronic, micro, optical and digital technologies to continually improve the features and functions of their cameras and create innovative discriminators against their competition to attract the technology-savvy consumers. Similar concepts are applied to other consumer and business products as well.

INNOVATION IN PROCESS DESIGN

Providing innovative product design and development processes that take products from concepts through to manufacture is imperative. Creative ideas and innovative solutions must focus on increased quality, speed of production, reduced production costs, increased product sales, and ease of maintenance.

Focusing innovation based on the actions of market leaders is disadvantageous because only the effect of their design efforts is visible. The intellectual property strategy, and product and process knowledge that produced these artifacts, are not only hidden but also typically one design-cycle old. Focusing solely on these artifacts is likely to perpetuate a market-following position.

A healthier strategy is to evaluate the design space by quickly assessing each area to determine its potential value, its value with current technology in the current marketplace, and the technology hurdles and market factors that must be overcome to yield its potential. This strategy allows a design team to focus on areas of the design space that present the most valuable innovation opportunities while understanding and managing the risk that competitors might pursue other areas. This strategy limits competitive risk and enables the business to react with agility when new technologies or market changes present opportunities.

As the technology revolutions occur, engineers generate innovative techniques to apply these technologies to process design improvement. The design techniques continue to evolve from traditional manual design using ad hoc approaches to automated design using Computer Aided Design (CAD) tools. Such CAD tools can help develop several variations of the conceptual design on the computer without using any product material. The conceptual designs can be evaluated for feasibility and usability and can be improved by applying variations before they are sent for manufacture.

The design techniques continue to be innovated from traditional unstructured design to structured design and then to the object-oriented design (OOD). A significant paradigm shift occurred when moving from structured design to OOD, as the structured design techniques were used for a very long time. The OOD approach established a modular approach, whereby individual components can be changed without affecting the entire product or solution.

The OOD approach helps engineers create standard and modular design patterns that can be applied quickly to develop a new product or solution rather than developing each new product from scratch. Thus, the OOD approach introduces reusability of components that helps reduce production cost, improves quality for reusing proven components, significantly reduces production time, and improves speed to market. Innovations in both industrial design and information technology design techniques continue to evolve toward improved and efficient production of quality products at comparatively lower cost.

Innovation in Business Growth

A business may produce the best products today and generate big demand. However, if that business does not continue to innovate to improve the product, its quality, functionality and service, it may soon be surpassed by the competitors who are continually innovative in order to improve on quality and functionality.

As an example, after the airline industry was deregulated, it faced intense competition from a cost and service perspective. Such competition encouraged the airline companies to develop innovative approaches to keep their loyal clients and attract new ones. One of the airlines started with an innovative idea of introducing the mileage credit award that could help retain and attract more clients. In the beginning, this innovation significantly helped the airline in growing its business. Soon, however, other airlines followed the same lead, and finally this innovative idea reached the upper saturation side of the S-curve. As a result, the airlines were again challenged to develop new innovative strategies, and thus the competitive race goes on. This example helps to illustrate that continuous innovations in business strategy are extremely important for the success and future growth of any company.

With the advancement of technologies, market trends and competition, the speed in which innovative ideas should be generated continues to increase. Any new ideas become obsolete quickly and reach the saturation points of the S-curve. Besides the speed of generating innovative ideas, ensuring that the new innovation focuses on generating unique discriminators, which can demonstrate better products compared to any competitive products, is even more important.

INNOVATION SUCCESSES

Why did Dell Computers succeed in entering the saturated personal computer marketplace in the 1990s, when this marketplace was so saturated and overwhelmed by large companies (IBM, HP and DEC) and a large number of clone PC producers everywhere possessing very low prices? The answer is this: Dell analyzed the PC marketplace, assessed its problems, issues and weaknesses, and determined new innovative strategies to overcome such problems and bottlenecks. Dell's strategies included offering build-to-order computers that provided the exact configuration the client required as well as managing to keep a very low inventory. Dell's strategy of direct selling to consumers eliminated distribution and retail costs and passed along additional cost-benefits to the clients. Such innovative ideas not only helped Dell succeed in entering the high-risk PC market, but they made Dell the leader in capturing the largest share of the market today, beating all the large and small competitors.

Apple Computers struggled for its survival in the 1990s after its founder, Steve Jobs, was fired by the board. Jobs had been the thinker and initiator of innovation behind Apple's success. Subsequently he returned to the company, and through a series of innovative ideas and risks, he returned the company to the path of success and growth. Apple is now a highly profitable company with a market capitalization of $38B and growing.

3M established one of its core values as an "Innovative and Practical Solutions" company that helps its employees always formulate new product ideas that are also practical in the marketplace. Sticky Notes™ is an example of a simple creative innovation that has tremendous practical use everywhere.

IBM has always focused on innovation. Its first CEO, Thomas Watson, Sr., introduced and promoted the THINK concept, where all employees had a sign of THINK at their desks that reminded them to always think and be creative in every task and activity. IBM's goal has always been to be the leader in the invention, development and manufacturing of the industry's most advanced information technologies, including computer systems, software, storage systems and microelectronics. In order to stay focused and achieve its key goal, IBM invests about 8–10% of its gross income on research and development to ensure its leadership in producing the most advanced information technology products and solutions. IBM holds the highest number of patents compared to any other company in the world.

DEMAND FOR INNOVATION

The importance of innovation for the economic performance of industrialized countries has been largely stressed recently by intense worldwide competition as the world becomes open, flat and global. In the last decade, as the nations of the world economy are becoming increasingly open and interdependent, innovation has become an even more important contributor to economic well-being. The intensity of innovation activity is very different across different countries. Several industrialized countries make significant investments in research and development and consistently receive numerous patents, whereas some of the countries make very little or no investment in innovation. If these two categories of countries are compared, the countries that invest in R&D and focus on innovation in their activities continue to enjoy healthier economic growth compared to the ones that are less innovative.

Innovation is an extremely powerful tool that can not only bring success in producing and marketing better quality products but can also help establish new opportunities and marketplaces. Because of intense competition, every business has pressure to be innovative in its approach in producing new products. Thus, each business demands innovation from its employees and processes and typically establishes a well-defined structure and organization to perform R&D and produce innovative solutions and results. This concept is referred to here as *Innovation on Demand* or *Demand Creating Innovation.*

Innovation Creating Demand is a complementary concept where some innovative research breakthroughs may create a totally new marketplace (as mentioned under "The World of Innovations" in Chapter 1). For example, the innovation of the World Wide Web concept created a totally new marketplace for all industry sectors, in that it has connected the world together so quickly beyond anyone's imagination and expectation. Before the invention of the Web, computers were used by expert programmers, engineers and trained personnel.

The Web innovation, however, established the new market where everyone, including grandparents, are hooked to the Internet to access such information databases as their bank accounts, receive grandchildren's pictures, send electronic messages to their dear ones (including birthday and Valentine's Day greetings), and interact with other vast information resources online without the assistance or intervention of any operator. Similarly, the innovative breakthrough in developing a semiconductor transistor by the Bell Labs scientists about half a century ago revolutionized the entire world of electronics and computing and established tremendous market opportunities in all sectors of businesses, from consumer products to heavy industrial products.

These two concepts can be compared with the famous "demand" and "supply" concepts as well, where supply must be increased to meet the demand. Similarly, businesses increase their supply of products and create demand to sell on discounted or volume prices.

IMPACT ON THE INNOVATION PROCESS

The demand for innovation continues to increase with growing and intense global competition and customers' ever-growing requirements and sophistication. The companies that establish processes and investments that continually improve the functionality and quality of products, enhance the design and manufacturing processes, and establish creative marketing strategies succeed in increasing their market share and achieving sound business growth. However, the R&D processes and organizations that are established for generating innovation should be periodically evaluated for their efficiency and effectiveness to ensure that they continue to generate productive innovations for achieving sound business results.

Innovation processes must continue to evolve and improve to stay tuned with the demand, competition, latest technologies and market trends. The R&D team should never spend too much time at the saturation point of the S-curve. The processes and organization focusing on *Innovation on Demand,* and the results of these activities, must be continuously monitored, enhanced and improved to ensure innovation effectiveness and successful business results. The R&D team should be provided with on-going training and education to keep members up-to-date with other innovations, competitive products and the latest technology trends. They should also be provided with up-to-date facilities, equipment and environments to be productive in their innovation activities.

TAKE AWAY

1. The organization should encourage two types of creative activity for its employees: a) *lateral thinking* will help develop a list of alternatives and options and b) *vertical thinking* will help select the best alternative among all the selected alternatives.
2. In an organization, structured and disciplined generation of innovative ideas is needed at various levels of activities—from business strategy, to solving client problems, to inventing and implementing new and improved products, solutions and services.
3. At the top of the S-curve, applying *out-of-the box* thinking to come up with new products or approaches is needed.

4. Various types of technologies—from *low-tech* to *high-tech* to *super-tech*—exist. As we move from low-tech to super-tech, more innovation is required.

5. If a business is involved in a super-tech project, it will demand more innovation in new ideas for new products and solutions to solve complex problems.

6. Ensuring that the new innovation focuses on generating unique discriminators, which can demonstrate better products compared to any competitive products, is even more important than is the speed of generating innovative ideas.

7. The importance of innovation for the economic performance of industrialized countries has been largely stressed recently by intense worldwide competition as the world becomes open, flat and global.

8. The countries that invest in R&D and focus on innovation in their activities continue to enjoy healthier economic growth compared to the ones that are less innovative.

9. Companies like Dell Computers, Sony, 3M, FedEx, Motorola, Nokia, IBM, Ford, Toyota, TATA, ABB, Reliance, Phillips, and BP AMCO have achieved success through innovation.

10. The *Innovation Creating Demand* concept represents R&D-driven, slow-paced innovation, while innovation on demand requires organization-wide, fast innovation.

11. The demand for innovation continues to increase with continuously growing and intense global competition and customers' ever-growing requirements and sophistication.

12. The companies that establish processes and investments that continually improve the functionality and quality of products, enhance the design and manufacturing processes, and establish creative marketing strategies, succeed in increasing their market share and achieving sound business growth.

13. The R&D processes and organizations that are established for generating innovation should themselves be periodically evaluated for their efficiency and effectiveness to ensure that they continue to generate productive innovations for achieving sound business results.

PART II

Understanding
Innovation

BRAIN HARDWARE AND INNOVATION PROCESSES

PRAVEEN GUPTA

As we move towards a better understanding of the innovation process, we must begin with an appreciation of the primary source of innovation—the human brain. In order to drive an automobile, one does not need to know the details of all its components. However, a general understanding of the major functions and controls helps in optimizing the performance. This chapter gets into the details of how the brain functions and provides an overview of those brain processes that are related to innovation.

INNOVATION AND THE BRAIN

This chapter introduces the need for developing a greater understanding of the brain in order to accelerate innovation. The growing field of neuroscience has been examining more about the brain. A total understanding of the brain may be difficult to achieve; however, understanding the innovation process at a higher level may be a significant contribution to improving intellectual productivity in the business environment. This chapter at best creates a crude framework for raising awareness of the brain's role in institutionalizing innovation in organizations. Innovation does occur in the brain somewhere, appearing to be random because of its frequency of occurrence. A little more understanding of the brain processes can lead to a huge improvement in the rate of innovation, which is needed in the knowledge age.

Innovation has been considered an art for centuries. Innovators, however, believe they know how to repeatedly engage in it. For them, innovation is a science. Some people have dozens or even hundreds of officially-granted patents. For them, they have mastered the art of innovation, or at least they have developed a science to their art, just like Thomas Edison, who decided to innovate and had a goal to file one patent per week. Edison ended up with over 1000 patents in his name. For Edison, innovation was not an art; it became a science. What did these innovators do that can be learned by others?

Einstein's brain, though a little larger than normal, happened to be within acceptable statistical variations observed among human brains. Today, extensive work in the area of neuroscience is aimed at improving human health and building more intelligent machines for doing tasks in a similar way as humans do them. Businesses have stretched the physical limits of employees by improving productivity through tools such as computers and automation. As a result, however, people generate a lot of information without effectively using it.

The new methodologies such as Six Sigma require newer solutions faster. The required rate of improvement has been taken to a new and higher level. Sustaining growth and realizing profits require thinking differently. "Thinking" is the hardest task to do, though it should not be, as it is the core competency of humans when compared with other species in the environment. Thus understanding the brain and its functioning in the context of business innovation helps to accentuate usage of this practically-unused faculty. Any incremental increase in the average utilization of the brain will cause tremendous change and innovation beyond the imagination in every aspect of life.

Innovation begins with a thought. An innovative thought requires focused thinking and capturing many thoughts as the ideas move through activity, process, product and business stages. The brain is a building block of innovation, so understanding its functions, and increasing its utilization for accelerating innovation, can only help to further its innovative capability. Applying thinking to the innovation process, which is a somewhat unique approach to understanding it, is the first step in developing innovative solutions on demand.

As discussed in Chapter 4, several national policies have developed based on the assumption that clusters of innovation (i.e., groupings of people) lead to increased innovation. Clustering is a process of exchanging information. In today's interconnected world, everyone is practically connected and associated with many clusters. Thus the *physical* clusters of innovation no longer make sense. Instead what really happens in the cluster is the question that demands an explanation.

What is seen in the universe is what is seen in the mind. Every idea or discovery occurs in someone's mind somewhere. In some respects, the human brain is almost as big as the universe itself (i.e., has almost as much capability or potential). In fact, each and every brain is equally powerful within some variation. The utilization of intellectual capacity can be increased by understanding the brain's involvement in the discovery process.

The brain contains about 10 billion neurons (or nerve cells) and trillions of axons that facilitate connections with other neurons. Each neuron can have up to 10,000 connections (or synapses). The total number of possible neuron combinations is an astronomically high number (about 10^{80} connections). This number is equal to the number of positively-charged particles in the whole known universe (Restak, 1995).

The smartest person is similar to Einstein, whose brain (people assume) had 100% of the connections made. A "dumb" individual can be considered someone with only 1% of the brain connections made. Yet even that 1% of brain connections has practically infinite potential. Many great innovators were explicitly deficient in some aspects—some physically and others mentally.

Every human can innovate and must understand the potential to innovate. This chapter, through mostly an overview of the brain and its correlation with the innovation process, demonstrates that the innovation process can be learned and must be learned to realize a person's innovative potential, compete through discovering new products and services, and assist in sustaining profitable growth.

OVERVIEW OF BRAIN ANATOMY

The human brain starts forming within the first thirty days after the conception of a human being. The typical brain looks something similar to the schematic shown in Figure 6.1. The outermost layer (i.e., the hard surface similar to leather) is called dura mater and protects the brain. Below dura mater is cerebrospinal fluid, which provides a soft protective cushion.

The inner layer is called pia mater and is similar to a crumbled sheet of crepe paper, which if extended is the size of a dinner table napkin. This sheet, which is called the cortex, consists of gyri and sulci and is divided into two asymmetrical halves called the left hemisphere and right hemisphere (see Figure 6.2). Interestingly, the right side of the brain controls the left side of the body, and the left side of the brain controls the right side of the body. The two hemispheric sides of the cortex are connected through nerve fibers called the corpus callosum. Each hemisphere consists of four lobes, namely frontal (motor activities), parietal (image and recognition), occipital (vision), and temporal (hearing or timing). Each hemisphere, when viewed from the side, is divided in three sections named the forebrain, the midbrain and the hindbrain.

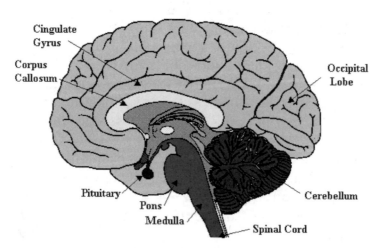

FIGURE 6.1 The Brain - Right Down the Middle [Reproduced with permission: Dr. Eric Chudler]

Left Brain Functions	Right Brain Functions
Sequential Analysis	Holistic Functioning
Logical Interpretation	Comprehension of simultaneous multi-sensory input
Language, Mathematics	Visual and spatial capability
Reasoning	Coordinated complex functions such as dancing, singing, and gymnastics
Language Memory	Visual, spatial and auditory memory

FIGURE 6.2 Brain Hemispheres

The frontal lobe takes up half of the hemisphere, controls motor activities and integrates emotion to convert thoughts into actions. The parietal lobe is the communication center for receiving and integrating all sensory inputs for creating information. The temporal lobe consists of amygdala and hippocampus that are involved in learning, memory and expression. The occipital lobe (the farthest of all) processes the visual information.

The human nerve system includes central and peripheral elements. The brain and spinal cord form the central nervous system. The peripheral nerve system communicates to and with the central nerve system. In the central nervous system, the brain's three elements of hindbrain, midbrain, and forebrain are formed within a few days after conception.

The hindbrain produces the medulla, the cerebellum and the pons. The medulla and pons are involved in controlling the physiological functions, like breathing or motor skills. The medulla is involved in the control of blood pressure, heart rate and breathing and is located right above the spinal cord. The cerebellum is the interface between the higher brain and the muscles. The pons connects the hindbrain with the midbrain and forebrain.

The forebrain develops structures like the diencephalon surrounded by the telencephalon (see Figure 6.3). The diencephalon is the core of the forebrain, consisting of the thalamus and hypothalamus. The thalamus is responsible for managing the sensory information going to the telencephelon, and the

Brain Modules	Nucleus Structures	Key Functions
Telencephalon	Cerebral Cortex, Amygdala, Hippocampus, Basal Ganglia	Analyze sensory data, perform memory functions, learn new info, form thoughts and make decisions; plays a role in sense of smell, motivation and emotional behavior; two hippocampi, located on both sides of brain, involved in anger, fear, pleasure, formation and retention of memory for facts (database), and are motivators for problem solving; initiation and direction of voluntary movement, balance movement,...
Dienephalon	Thalamus, Hypothalamus	Receives and integrates sensory inputs, and relays to cerebral cortex; control of body temperature, emotions, thirst, hunger,....
Mesencephalon	Substantia Nigra, Central Gray, Red Nuclieus	Transmitter of information, changes in metabolism, intellectual abilities including memory, judgment, abstract thinking; Spinal cord control, and auditory reflexes; Coordination of the instincts and emotions, reactions, and motion
Metencephalon	Pontine, Deep Cerebellar	Rapid Eye Movement (REM) sleep, subconscious thinking, dreaming, imagination, memory consolidation, comprehension; planning, predicting, and motor control
Myencephalon	Inferior Olive	Communicates sensory information and inputs from other brain nuclei and communicates sensory information to cerebellum

FIGURE 6.3 Brain Functions [Reproduced with permission: Dr. Eric Chudler]

hypothalamus is responsible for regulating physiological and biological functions. The telencephalon, or cerebrum, consists of the two cerebral hemispheres and is mainly responsible for sensory perception, learning, memory and conscious behavior (Purveys, 2004).

The telencephalon includes the structures such as the thalamus, hypothalamus, hippocampus and amygdala. The thalamus is located above the brain stem in the midbrain. The thalamus receives the information from various senses, transmits it to the cerebral cortex and other areas of the brain, and communicates signals from the cerebral cortex back to the spinal cord.

The hypothalamus acts like a thermostat to control the body temperature. If a person feels too hot, the hypothalamus sends signals to expand capillaries in the skin, causing blood to cool down. The hypothalamus is a small structure next to the thalamus and controls metabolic functions, body temperature, sexual arousal, thirst, hunger and biological rhythms.

The hippocampus surrounds the thalamus and is divided into more than one section. Each hippocampus is connected to a structure called the amygdale, which is responsible for fear and fear memory. Blocking protein synthesis of amygdale will block the formation of fear memory. The hippocampus and amygdala are responsible for processing and perceiving emotion and memory. The hippocampus evaluates or associates memories and forwards them from short-term memory to long-term memory in the cerebral cortex for permanent storage. Therefore, the hippocampus can be considered an important center of learning in the brain.

The grouping of the thalamus, hypothalamus, hippocampus and amygdala is called the limbic system, which deals with physiological drives, instincts and emotions. The limbic system is where the sensations of pleasure, pain and anger are most keenly felt.

The cerebral cortex is a sheet of gray matter that is about 4 ± 2 millimeters, or the height of 6 business cards (Hawkins, 2004), and makes up the outer layer of the brain. The cerebral cortex has bumps and grooves, normally called gyri or sulci (i.e., gyrus and sulcus for singular bump or groove), respectively. The cortex is full of neurons which are connected based on the information received throughout life. As a person learns, the neurons gain stimuli, grow dendrites, and connect with other neurons.

The cerebral cortex has several specific functional areas, and any area of the cerebral cortex not specified is used for association responsible for thoughts, judgment, humor and behaviors. The specific functional areas of the cerebral cortex include the prefrontal cortex (for problem solving and complex thoughts), the motor cortex (for bodily movement), the somatosensory cortex (for processing multi-sensory input), the visual cortex (for detecting simple visual input), and the auditory cortex (for detection of sound quality). As highlighted earlier, besides these functional areas, association areas in the cerebral cortex are responsible for advanced capabilities in various areas.

Basal ganglia located at the base of the cerebral hemispheres receive inputs from the cerebral cortex through striatum in the midbrain and forward it back to the cerebral cortex through the thalamus. Basically, the basal ganglia function to maintain the muscle tone needed to stabilize joints.

The brain stem controls basic functions of life, including muscle coordination. The brain stem in the midbrain processes visual and auditory information as well as the information between the higher brain and the spinal cord. The sensory information moving up the spinal system passes through the brain stem in the hindbrain area, where synapses are formed with brain stem neurons, called the reticular system. The reticular system controls bodily functions such as movement, coordination and sensitivity to pain. Activity level in the reticular system affects the nerve systems, thus affecting the awakening or sleep state of the brain.

The cerebellum controls fine body movement, balance and posture. The cerebellum is like a little brain inside the brain with its own cortex and even its own hemispheres. The cerebellum, though only about 10% of brain volume, contains about 50% of all the neurons in the brain (Best, 2005). The cerebellum receives commands from the somatomotor cerebral cortex and gives smoothness of motion and exactness of positioning through internal and external feedback comparisons.

THE INNOVATION PROCESS IN THE BRAIN

Medical researchers have mapped the brain inside out, and most of the functions are compartmentalized. The brain includes practically infinite numbers of neurons, which get excited by stimuli through senses and produce electric potential. This electric potential is then converted into chemical reactions that transform into various proteins and other reactions. The role of $Na+$, $K+$ and $Ca+$ ions in the functioning of neurons and the creation of synapses is understood; however, what happens next is difficult to comprehend. While some sporadic phenomena can be explained, the root causes of those phenomena are often unknown.

In an attempt to understand the innovation process inside the brain, identifying certain parts associated with language and auditory interpretation is possible, but exactly how a person thinks and makes decisions is a mystery. However, everything that a person sees, does or imagines does happen inside the brain. Actually, the brain makes it possible to have some knowledge of everything in the universe. If a person knew how to use the brain effectively, then everything discovered so far could have been discovered by one brain (given the recognized size of a person's brain).

The total absence of the brain occurs when the brain is absolutely nonfunctional or dead. Otherwise, the brain is so big in its potential, that significant damage would not render it dead. For example, the brain may have approximately billions of neurons, trillions of axons, and many more combinations of neurons and axons, making synapses. Synapses can be considered the information bits. The number of synapses can realistically approach infinity. A fraction of that capacity appears to be 10^{80}. A small fraction of such brain capacity is still enough to use the brain throughout life (or for thinking of innovative solutions).

Einstein said that nothing is invented; instead everything is discovered. If everything is discovered, the brain must see everything. Just how can a brain see everything? In other words, what is the process used by the brain to try out different combinations to produce innovative solutions? The brain is continually bombarded with information. What happens to it?

Look at electrical connections as an illustration. If a connection is broken, the circuit is literally dead. If the connection is made, the room is lighted and everyone is awakened. In the same way, the brain appears to be a set of connections that must be present in order to sustain proper functioning. Any brain function is a result of cumulative output of the input stimuli, be it visual, audio, touch or taste.

Figure 6.4 shows a simplified version of the brain process. The cerebral cortex appears to be a collection of nerve cells called neurons. Brain neurons are available in practically limitless quantities to absorb sensory input for processing. Signals are received through various senses or activities. These signals are passed through nerve fibers to neurons in the cortex area for pattern formation.

Once the pattern is formed based on the purpose to use it or not to use it, the information that is compared in the association is in the cortex and is stored in its short-term or long-term memory bank. The cortex area and surrounding gray

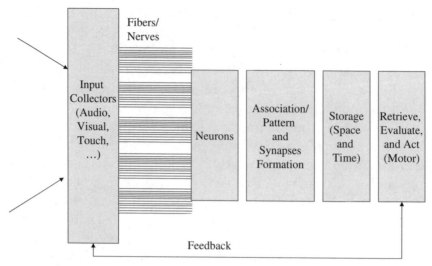

FIGURE 6.4 Mental Process Flow

matter can absorb short-term or long-term information in unlimited amounts. The short-term memory connections are elastic and reversible, while the long-term memory synapses are hardwired to last longer. Some patterns are permanently etched based on the extent of their strength, leading to virtually permanent connections or patterns. When the patterns are formed, they are evaluated or compared for signal generation and motor action.

The duration of the information storage is dependent on the elasticity of the material and the decision made to ignore or receive the material. The natural loss of information can be attributed to the property of the cortex material; however, how the decision is made in the brain to use or not to use the information is still unknown. Once the decision is made to use the information, processing of the information can be understood inside the brain. Humans have the capability to make such decisions without knowing the mechanism at this stage. Perhaps further research will reveal more secrets of the brain.

Once the information is imprinted on a group of neurons, a pattern is formed. The human brain continually forms patterns of signals it receives from various senses. Then the decision to recognize patterns for comparisons is made. The question remaining is this: Which part of the brain makes this comparison and how?

The brain contains visual and auditory (i.e., somatosensory) association areas where the association among patterns is made. If the match is made (in the context to use this information), an electrical signal is generated, which leads to a chemical reaction producing certain tactical or facial movements or actions. When innovation is taking place, the existing patterns are reviewed in the memory. To discover a new or unique pattern, comparisons (through association) of existing patterns must occur. Then the person making those comparisons must determine the missing patterns.

Figure 6.5 represents implementation of the innovation process through brain functions. For example, gathering information through sensory inputs is processed through the thalamus and hypothalamus. The repetition and intensity of input determines the chemical signal strength in terms of its elements and temperature. The combination of input type, intensity and temperature creates certain signals that determine people's

Elements of Innovation	Key Side	Key Brain Elements ...Accelerators
Gathering information	L	**Somasatory sensors, Thalamus, Hypothalamus** Receives and integrates sensory inputs, and relays to cerebral cortex, ...**Energy and comfort**
Learning, Comprehension	R	**Cerebral Cortex, Amygdala, Hippocampus** Analyze sensory data, perform memory functions, learn new info, form thoughts and make decisions,... **Motivation and Incentives**
Analysis, Questioning, Interpretation	L	**Substantia Nigra, Central Gray, Red Nuclieus** Transmitter of information, intellectual abilities including memory, judgment, abstract thinking, ... **Stamina and Time Management**
Association, Induction, Deduction	R	**Hippocampus** involved in formation and retention of declarative, spatial, or long-term memory for facts (database),... **Knowledge and Research**
Combinatorial Processing	L	**Stratium** responsible for procedural or short-term memory,... **Experiment and Play**
Extrapolation	R	**Pontine, Deep Cerebellar** causing subconscious thinking, dreaming, imagination, memory consolidation, comprehension,...**Rest and Reflect**
Formulation	LR	**Cerebellum, Ganglia** – Timing relationships, motor planning, predicting, and motor control,...**Evaluate and Select**

FIGURE 6.5 Innovation and Brain Functions

behaviors through their tactical or facial expressions. The signal strength depends upon the familiarity of the pattern, its repetition, and its intensity. When the patterns are associated or matched, a positive signal is generated. When the patterns are not recognized or missed, a negative signal is released.

Accordingly, the innovation process becomes a process of reviewing new or selected patterns in the cerebral cortex, evaluating them with existing patterns or potential patterns, and determining the missing or absent patterns. Such a process leads to signal strength for either generating ideas or questions. Thus, a thought or an idea is a feedback/observation of a pattern that is either present or absent in the cortex and then represented (or communicated) through tactile or facial expressions. Similarly, a question is an idea that is based on the nature of the resultant signal strength sent to Broca's and Wernicke's areas for asking questions. To ask questions, a person must first develop, understand and speak the language. Without the development of language and speech faculties, a question is most likely going to be suppressed.

Once the generation of thoughts is understood, questions about the speed of thought may be asked. With the current understanding of brain functioning, the speed of thought thus becomes the rate at which patterns can be associated and compared in order to evaluate and generate new thoughts or new patterns. Given the number of neurons, axons and synapses in the cortex, algorithms must be developed to speed up the evaluation process.

One way to speed up the speed of thought is to establish anchor locations in the brain. Anchors are long-term memory locations that have almost permanently-etched information. These locations are formed through repetitive use of the known pattern, which creates strong synapses that appear similar to cured synapses losing their elasticity. For example, children memorize tables that help them to perform mental math. Having the anchors in the memory, calculations are then expedited. Similarly, if certain patterns are anchored in the cortex, comparing patterns by association can be expedited. However, they may eventually be lost over a longer period.

ACCELERATING THE INNOVATION PROCESS IN THE BRAIN

Having understood the basic functioning of the brain, how to accelerate the innovation process becomes the next question. Acceleration can be achieved through creating patterns faster, thus speeding up the associations and evaluations. The creation of faster patterns requires continually gaining new experiences through somatosensory learning, as well as getting involved in more activities leading to increased input through the spinal cord. Creating associations and evaluating patterns faster requires anchors and the involvement of more neurons. These anchors and increased neural involvement can be achieved through speeding up thought and through external excitement of neural activities.

Figure 6.5 captures the activity and ambiance levels for improving various stages of the innovation process. For example, in order to improve information-gathering through sensory inputs, enough energy and comfort must exist. To improve the analysis and questioning aspects of the innovation process, stamina and time management are required; to improve the association and evaluation processes, knowledge and rest to reflect or think are needed.

External excitement implies application of an external energy to stimulate certain brain structures associated with the innovation process. Accordingly, studies have been done to map the brain's wave pattern in its different states. Taylor Andrew Wilson, for example, has utilized the Brain Wave theory to create music that accelerates the brain's performance. In his book *The Mind Accelerator*, Wilson identifies the Beta state (13–40 Hertz) as associated with an attentive, conscious, and narrowly-focused state of mind. The Alpha state (7–12 Hertz) is associated with states of mind involving visualization, relaxation and ingenuity. The Theta state (4–7 Hertz) is associated with intuition, memory, and deep-thought states of mind.

According to Wilson's theory, to sustain various activities in the brain, music with certain brain waves is applied. For example, to investigate potential combinations or patterns, the selected brain-wave music will bring brain waves into a peak-performance frequency, thus capitalizing on the information-acquisition and knowledge-building efforts. To expand thinking in order to explore potential innovative solutions for generating significant change, therefore, the selected brain-wave music must trigger optimal waves of neural firing. Using Wilson's theory, such waves will thus deliver a focused and acute peak-performance brain state for evaluating and analyzing ideas, suggestions and solutions.

Proper diet and exercise in order to fully utilize brain potential cannot be under-stated. Brain activity can actually be accelerated by increasing neuron activity level through proper diet and exercise. Proper nutrition and exercise helps to maintain the chemical balance of the Na^+, K^+ and Ca^+ levels needed for reaching maximum potential in innovation activities.

TAKE AWAY

1. One must understand that innovation does occur in the brain somewhere. A better understanding of the brain processes related to innovation can lead more people to believe in their innovative potential.
2. Innovation begins with a thought. "Thinking" is the hardest thing to do, though it should not be as it is the core competency of humans when compared with other species in the environment.
3. Any incremental increase in the average utilization of the brain has tremendous change potential and innovation beyond imagination.
4. A brain contains about 10 billion neurons or the nerve cells and trillions of axons that facilitate connections with other neurons. Each neuron can have up to 10,000 connections, called synapses.
5. Analysis of the brain functions suggests that both hemispheres are important in developing innovative solutions.
6. Organizations must establish stronger incentives for learning rather than focusing on the outcomes alone. Learning minds tend to be more innovative.

7. Organizations must give employees time to think; otherwise, a valuable resource is wasted.

8. Given the recognized size of a person's brain, everything discovered so far could have been discovered by one brain, but only if one knew how to use the brain effectively.

9. The speed of thought affects the rate at which patterns can be associated and compared in order to evaluate and generate new thoughts or new patterns.

10. The creation of faster patterns requires continually gaining new experiences through somatosensory learning as well as getting involved in more activities leading to increased input through the spinal cord.

11. To improve the analysis and questioning aspects of the innovation process, stamina and time management are required; to improve the association and evaluation processes, knowledge and rest to reflect or think are essential.

12. Another way to accelerate brain activity is to increase the neuron activity level through proper diet and exercise in order to maintain chemical balance, including the Na+ (Sodium), K+ (Potassium) and Ca+ (Calcium) levels.

FRAMEWORK FOR INNOVATION

Praveen Gupta

Innovation is a function of resources dedicated to it and the speed of thought—which itself depends on knowledge, options and imaginative talent. In addition, innovation can be categorized as Fundamental, Platform, Derivative and Variation. Depending on the nature of innovation desired, the combination and proportion of input elements can vary. A structured approach to breakthrough innovation, termed Brinnovation, is proposed. The understanding of these concepts should allow any individual or corporation to be more innovative.

INTRODUCTION

Striving for immortality or comfortable living drives the human appetite to innovate better ways of living. The innovation can be in drugs, foods, tools, communication or even astronomy. The drive to innovate originates from a fear of extinction at the extreme, or out of the need to make life more comfortable or free from suffering. The focus of successful innovation is to fulfill human needs, which may be health, food, work, communication, security and knowledge. Any drug for longevity, delicious food for better health, instruments for earning a salary, devices to communicate faster, weapons for protection, and methods of knowledge acquisition to perpetuate an appetite for more will lead to successful innovations, if they are affordable.

Peter Drucker (2002) observed that the innovation must be purposeful and begin with an analysis of the opportunities. Accordingly, the innovation must also be simple and capable of performing at least one specific task. Drucker identifies seven sources of innovation that include a flash of genius, exploiting incongruity or contradiction, growth in demand, changes in demographics or perceptions, and creating new knowledge. Recent knowledge innovations include history-making innovations such as the personal computer, cellular phone, iPod, the hybrid, and the Internet; older innovations include the airplane, wireless technology, electricity, cement, penicillin, and many more. Finally, innovations include the discoveries in space and materials, such as finding new planets like Neptune or discovering new elements such as Uranium. These innovations conventionally require commitment, hard work and perseverance.

As mentioned in Chapter 1, at the early stages of the discovery of electricity, corporations used to have Chief Electricity Officers; in the information age, businesses appointed Chief Information Officers; now in the knowledge age, businesses are appointing Innovation Officers more frequently. Innovation is moving towards becoming a standard process similar to other functions in business, such as purchasing, sales or quality control. Becoming a standard process implies a standard task must be performed; the outcome is predictable to some extent; personnel for the process are designated; a box on the organization chart exists; and a room for standard processes is labeled in the facility.

Similarly, as the innovation process moves toward becoming a standard process, people must designate a room for innovation; personnel for innovation must be present; an innovation box must exist on the organization chart; and someone must assume the role of Chief Innovation Officer. Some companies already have appointed innovation chiefs, including Coca-Cola of USA, DSM of Netherlands, the Health Science Centre in Canada, Publicis Groupe Media of France, and Mitsubishi and Hitachi of Japan, implying their focus and resource commitment to innovation in sustaining profitable growth.

Companies like 3M, Proctor & Gamble, AT&T, IBM, Siemens, Sony, Toshiba, Airbus, Unilever, Ford, GM, Tata, and Birla have been innovating for many years. Even large corporations, however, are realizing that the innovation process utilized thus far may not be as effective competitively as it was in the past. IBM offers Innovation on Demand consulting services to other businesses, yet even experts at IBM are realizing that their current understanding of innovation must improve.

In 2004, IBM organized a Global Innovation Outlook (GIO), which initiated a global dialogue for learning and changing the nature of innovation. Participants represented academia, the government, non-government organizations, corporations, venture capital firms, think tanks and experts. The GIO reports that up until now, business must have mistaken invention for innovation. One of the obvious changes recently is that innovation is occurring faster and taking a shorter amount of time.

Participants of the GIO consequently recognized the need to redefine innovation. Accordingly, the consensus opinion was that "We must define 21st century innovation as beginning at the intersection of invention and insight: We innovate when a new thought, business model, or service actually changes society." This redefinition of innovation demonstrates that businesses must adjust their understanding of innovation in the knowledge age.

When comparing innovation in the 20th century with the 21st century, innovation in the 20th century was a forte of large corporations with tremendous resources for research and development. The smaller corporations followed their lead by developing some derivative products. New knowledge had protection from reuse without paying royalty for it. Large corporations grew and employed thousands of people. The standard of living improved and families saved money. At some point when risk was manageable, entrepreneurs tried something new related to their work at larger corporations, and the idea spun off. New companies formed and became larger corporations. The new, large corporations did not fund the basic research and development as their predecessors did, however, because they could not afford to continue to do the research.

In the 21st century, however, knowledge acquisition is now a decentralized process due to the invention of the Internet. People have access to knowledge everywhere there is access to the Internet. With the Internet, the control of knowledge has fragmented all the way down to individuals. As a result, the rate of innovation is changing, and large corporations cannot keep up with it. Many new companies start up with new ideas funded by the resources of venture capital firms. Therefore, larger corporations must recognize the new model of innovation. Some of the large firms, like Proctor & Gamble, have set goals to seek a certain percentage of innovation from outside the corporation's boundaries. Such outsourcing of innovation is called "open innovation." *example ? real time:*

Examining the last century highlights the changes that have occurred from horse cart to space shuttle, labor to automation, material flow to information flow, and physical resources to intellectual resources. For example, the physical resources included time and material, while the intellectual resources imply knowledge. Figure 7.1 shows a comparison of innovation in the age of time and material with the knowledge age in terms of material, machines, methods, people skills, testing instruments, and the environment.

For the innovation process to be repeatable and available for any innovator, from the individual to larger corporations, some standard process must be established. In order for the innovation process to become repeatable, it must first be understood. To understand the innovation process, a basic framework must be discovered that can then be used to fill in the blanks for producing a repeatable innovation process. The

Key Resources	Time and Material/ Physical Age	Information/ Knowledge Age
Material	Raw material	Information
Tools	Machines and tools	*Brain (to be understood)*
Methods	Repeatable Processes for well understood machines	*Repeatable process to be developed*
People	Workers for physical effort	Workers with thinking effort
Environment	Comfortable for producing goods	Learning and creative
Expectation	High volume reproducible products	High volume customized solutions
Measurements	Productivity and Performance	Performance and productivity

FIGURE 7.1 Ages of Innovation

process must work with any brain, be easy to understand, and be logical enough for many people to use repeatedly. As one of the professors working in a major university said, "We do not understand the innovation process well enough for it to be standardized."

In the knowledge age with access to the Internet, a networked individual is the building block of innovation. As shown in Figure 7.2, now the individual has access way beyond the network of a few individuals; instead, practically the whole world is the network. Once the network really becomes ubiquitous, the individual becomes the building block of innovation with access to laboratories, universities, experts, corporations, or trademark and patent offices. This phenomenon has already begun; however, its full impact is yet to be realized.

Once the building block of innovation is established, defining the innovation is equally important. Creativity and innovation are interchangeably used; however, experts define innovation as the process where a creative idea is applied to produce value to society. The current process of innovation is from idea to commercialization. Until the idea becomes a success, it remains a creative idea. Once the creative idea becomes successful, however, it is a major breakthrough—an innovation.

In the knowledge age, utilizing the processes of both Einstein and Edison to produce knowledge solutions is needed.

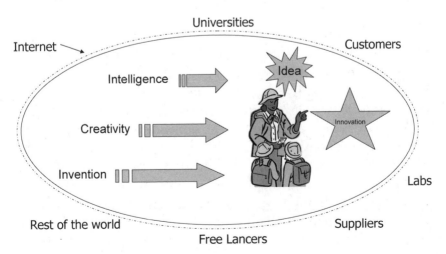

FIGURE 7.2 Building Block of Innovation

Einstein helped in learning about the theory of innovation, and Edison helped in understanding the methodology of solution development. Interestingly, Einstein did no physical experiments, while Edison had laboratories and the innovation room for developmental experiments. Einstein believed that every innovation is discovered, and Edison believed innovation could be produced on demand. Innovation thus becomes the discovery of an innovative solution on demand. With this understanding, a framework for innovation can be developed, and a methodology for innovation can be established.

INNOVATIVE THINKING

Einstein's Exhibits in the Boston Museum of Science show that most of his innovative work was published through four papers in 1905. Einstein tried to put the puzzle together with various pieces of nature, which he then tried to include in his unsuccessful effort in developing a unifying theory. Even with the availability of various innovation methodologies, tools and practices, a framework for innovative thinking is yet to be developed. Without such a framework, the predictability of methodologies, and the repeatability of the innovation process, cannot be established with confidence. The author has developed a model called Gupta's Einsteinian Theory of Innovation (GETI) to provide this needed framework for innovation.

GETI is based on Einstein's famous equation $E = mc^2$, where the "E" represents energy, the "m" represents mass, and the "c" represents the speed of light. Einstein's equation delineates the relationship inherent in the conversion between mass and energy. Every activity in nature is a conversion process.

Human beings are also energy converters. People consume resources and convert energy through actions or rest. Energy conversion can be physical or intellectual. The intellectual burning of energy occurs through thinking. When thinking, the ability to focus or channel energy into a direction, start associating various experiences with one another, and generate thoughts based upon those associations is critical. Thinking is continual, purposeful or inadvertent. When inadvertently thinking, people

think randomly, while when purposefully thinking, people channel their thinking into some direction to get an answer.   *example*

Thus, innovation begins with an idea, which is an outcome
of the thinking process; it must have some energy associated
with it. Sometimes a lot of energy is required to think of an
idea for some purpose. People even literally scratch their heads
to stir up and stimulate thinking. Thus, every idea must have
some energy associated with it that is an outcome of effort and
the speed of the thought.

All discoveries occur in the human brain. As shown in
Chapter 6, Brain Hardware and Mental Processing, certain
parts of the brain contribute to the innovation process. The
brain has a cortex that consists of billions of neurons (cells)
and trillions of axons (connectors). Neurons and axons form
connections called synapses. With billions of neurons and trillions of axons, the number of possible synapses is practically
infinite. The way the brain continually processes information is
by comparing stored and received information. The speed of
thoughts relates to how fast the brain processes the information stored and received, thus generating an idea.

For example, a patch of cortex consisting of 75×30 neurons
is shown in Figure 7.3. In order to match object A with an object
A′, the brain could compare one cell at a time to see if the objects
match. Such a process may take thousands of comparisons.
However, an experienced mind that has built anchors can quickly
hop from object A to object A′ and make the match. One such
comparison can take about five milliseconds. If the brain makes
thousands of comparisons, it can take hours, but if the brain
hops through the anchors, comparisons can happen quickly.

The numbers or matching time could be astronomical—
practically forever—if the person makes comparisons one cell
at a time. The speed of thought, however, can accelerate by
installing anchors based on multi-disciplinary experiences. The
actual brain speed can be one synapse per five milliseconds.
However, since the brain has billions of neurons and trillions
of axons, billions or trillions of synapses can form in parallel
within five milliseconds. Moreover, if the anchors already exist,
the speed can be faster.

As to the comparison with the speed of light, the speed of
thought can be much faster if the anchor is already established

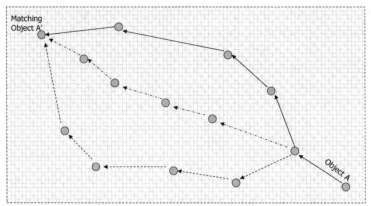

Example: Sample Cortex with 75x30 neurons ◉ => Anchor

FIGURE 7.3 Speed of Thinking

in the brain for two points irrespective of their distance. If a person has been on the moon or to a location light years away, the brain can hop to that location right away in the mind on demand. Interestingly, the capacity of the brain is practically infinite with respect to visualizing the universe. Richard Restak, M.D. (1995), in his book *Brainscapes*, mentioned that if the number of synapses is about 10^{80}, that number is considered to be the same as the number of atoms in the universe. The brain can handle many objects, perform associations, discover new things and innovate.

Thus, an innovation is a transformation of one set of ideas into another set of productive ideas. Therefore, the speed at which a person can process these thoughts becomes an important factor in accelerating or creating innovation on demand. Applying Einstein's equation to the process of innovation, one can equate "E" to the energy (value) associated with innovation, "m" to the physical effort or resources allocated to innovation, and "c" to the speed of thought, which can be faster than the speed of light. Restating Einstein's equation with proper substitutions, GETI delineates the following relationship:

*Innovation Value = Resources * (Speed of Thought)²,*

where the speed of thought can be described by the following relationship:

Speed of Thought ≡ Function (Knowledge, Play, Imagination)

The units of the Innovation Value can be represented in terms of resources and ideas over the unit of time, which can be equated to a new unit, Einstein (E), with the maximum value of '1.' Thus, the innovation value can be increased with more resources or faster generation and processing of ideas. The innovation value accelerates with better utilization of intellectual resources rather than merely allocating more physical resources to innovation.

The following matrix defines various terms and demonstrates an example of the quantification of innovation:

MATRIX FOR USING GETI FOR ASSESSING PERSONAL INNOVATION

RESOURCES (R)	KNOWLEDGE (K)	PLAY (P)	IMAGINATION (I)	INNOVATION VALUE (Iv)	COMMENTS
Degree of resources or time committed	Extent of knowledge based on research and experience	Percentage (%) of possible combinations of various variables explored	Dimension extrapolated as a percentage of ideal solution for breakthrough improvement	Estimated Innovation Level	This is an initial estimation of the proposed model. Further work is required.
50% (Limited time and insufficient resources)	75% (Significant knowledge and experience gained, some latest work is to be explored)	40% (Percentage of combination of variables explored mentally, experimentally or through simulation. Work is in progress)	66% (Selected dimension is extrapolated such that improvement is expected to be about 30%, which is about 66% of the breakthrough improvement)	0.182 (Long way to find an innovative solution due to lack of effort and play. To accelerate, one needs to improve all elements of innovation.)	Innovation Value = 0.5* ((0.75 + 0.4 + 0.66)/3)² = **0.182 Einstein**

Note: In the absence of a fully developed relationship among the variables, an additive relationship has been used to determine the Innovation value. (Gupta, 2005)

In other words, the Innovation value is equal to the Resources (commitment) times a function of knowledge, play, and imagination (KPI) squared. More than its numerical value, the equation identifies elements of innovation in order to maximize the innovation value. Most innovations are based on research, current experiments and innovative thinking. Measuring knowledge and quantifying combinatorial play are possible, but measuring imagination is difficult due to the complexity of mental processes. Therefore, imagination is transformed in quantifiable terms by understanding that *pure imagination is a random extrapolation*. Thus, imagination becomes a measurable component by nature of extrapolation.

INNOVATION CATEGORIES

Reviewing contributions of great innovators, specifically Einstein, Galileo, and Edison, Einstein engaged in mostly theoretical innovation, Edison innovated practical or business solutions, and Galileo did a combination of both. Einstein's work was fundamental in nature, while Edison's work was more tangible. Einstein conducted mostly thought experiments, like riding the light wave, while Edison conducted his real experiments in his laboratory. Looking at various innovations, they can be classified into four categories based on the amount of effort and the speed-of-thought component. The four categories of innovations are the following:

1. Fundamental
2. Platform
3. Derivative
4. Variation

The **Fundamental** innovation is a creative idea that leads to revolution in thinking. Such innovations are based on extensive research and are extremely knowledge driven, are theoretically proven, and lead to follow-up research and development. Such innovations occur with the collaborations of academia, commercial laboratories and even corporations. The fundamental innovations may lead to changes in thinking, extend an existing theory, or be a breakthrough concept with enormous impact, perhaps leading to the evolution of a new industry.

Actually, such innovations contribute to human evolution. Examples of such innovations could be Einstein's Theory of Relativity, Light Quanta or photon, electricity, penicillin, the telephone, Xerox, wireless communication, the transistor, computer software, UNIX, the Internet, the Fractal, the Edison effect and planes. The fundamental innovation has a significant academic component of science, which makes it available for the common good and thus less commercially protected.

The **Platform** innovation is defined as one that leads to the practical application of fundamental innovations. Such innovations normally are launching pads for a new industry. Examples of platform innovations include personal computers, silicon chips, cell phones, digital printers, Web technology, Microsoft Windows, Databases, CDMA, Linux, drug delivery devices, satellites and the space shuttle. The platform component increases the portion of the laboratory or development component more than do fundamental innovations. The platform innovations launch industries, change people's way of living, and fulfill the basic purpose of innovation, which is to live longer more comfortably.

The **Derivative** innovation is a secondary product or service derived from the platform innovation. Derivative innovations include new server-client configurations based on the new network architecture or operating system for a cell phone, for example. Derivative innovations are slight modifications of the main product. In the case of Microsoft-like software, the platform is Windows, and derivatives are a new office suite; for CDMA-like platforms, derivative innovations are various features available to service providers; for a major satellite system, the derivative innovations are various launching options or capabilities offered to users.

The **Variation** innovation is the tertiary level of innovation that requires much less time and is a slight variation of the next-level products or services based on the derivative innovations. For example, variation innovations in cell phones are various color covers, ring tones, camera features, and more software-based optional features. In the case of Microsoft software, variation innovations are various applications developed and based on the Microsoft platform and derivative innovations. Typically, the Variation innovation occurs close to the customer and may be the candidate for reaching the ultimate in speed of innovation or innovation on demand in real time.

Understanding types of innovations and their relevance to a business helps in establishing appropriate goals for innovations and devising correct measures of innovation. Figure 7.4, Attributes of Innovation, lists various aspects of innovation. Innovation on demand can mean different things to different levels of innovation.

Over time, responsibilities regarding where to put resources and who can reposition such resources must be understood. For example, switching systems, chip manufacturing facilities, and basic material or technology research have gone beyond the affordability of businesses; their collaboration with one another, or the government's support, will come into the picture in order to further fundamental or platform innovation. Based on the commercial success of an innovation, the innovation can move to the next level up or higher. For example, a cell phone like Razor (Motorola) becoming so successful can

Types of Innovation	Primary Drivers	Key Aspects	Deliverables	Frequency	Time to Innovate	Ownership
Fundamental	University/ Laboratories	Science/ Knowledge	Concepts/ Revelations	Rare	Years – Months	Govt. (s)
Platform	Corporate R&D	Technology/ Large Sys.	Equipment/ Capability	Sporadic	Months – Weeks	Govt./ Business
Derivatives	In-house/ Outsourced	Application/ Small Sys.	Product/ Service	Regular	Weeks – Days	Business/ Individuals
Variations	Networks/ Individuals	Disposables/ Ideas	Packaging/ Integration	Continuous	Days – On-demand	Individuals

FIGURE 7.4 Attributes of Innovation

become a platform in itself (rather than a derivative innovation of a larger strategy). Microsoft Office is a platform innovation based on its success, and many additional, diverse, next-tier products or applications may be developed.

Various types of innovations are achieved by differing degrees of the speed of thought. For example, a fundamental innovation may require a more meditated process to think of theories, concepts, or solutions without major experimentation. In fundamental innovation, knowledge and imagination are key components. Interestingly, most of Einstein's work was completed in his mind rather than in a laboratory. He typically conducted "thought experiments."

The platform innovation involves relatively less knowledge and imagination and more play or experimentation. Figure 7.5, Speed of Thought vs. Type of Innovation, shows that variation innovation requires more play or more development effort than research and reflection. The chart helps in understanding how various innovators focus on a particular area for their work, as well as how various innovations are achievable by focusing on the right component of speed of thought. Actually, speed of thought becomes the manifestation of the speed (or rate) of innovation.

Figure 7.6, Extent of Innovation, shows that fundamental innovations can take a much longer time than do the variation innovations. More variation innovations will result than will

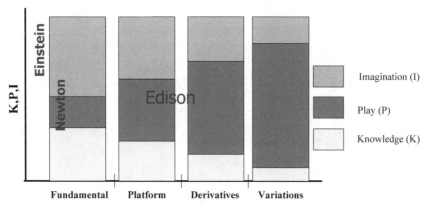

FIGURE 7.5 Speed of Thought vs. Type of Innovation

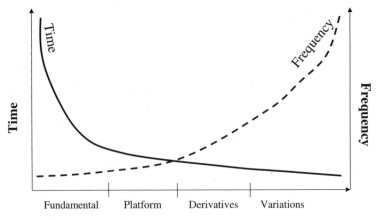

FIGURE 7.6 Extent of Innovation

fundamental innovations. The fundamental innovation is a rarity, while the variation innovations are continuously occurring.

Figure 7.7 graphically depicts the innovation process, which appears to be linear at first glance. However, any step within the linear process has nested loops or divergence. As an overall process, the innovation process must be streamlined and appear linear in order to show progress. The innovation process is based upon the innovation framework and is designed to produce breakthrough innovation (Brinnovation™) on demand. In

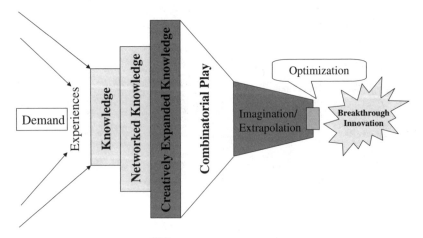

FIGURE 7.7 Brinnovation(TM)

other words, the innovation begins with a demand, and it must be purposeful.

THE BREAKTHROUGH INNOVATION PROCESS

The first step of the innovation process is to listen to the requirements for innovation and then gather knowledge about the topic in order to identify necessary inputs to the innovation. The networked individual, or the innovator, gathers more knowledge to achieve a certain level of competency in the field quickly. At this stage, process thinking helps in identifying input for the intended innovative solution. This step is a critical one and differs from the current methods of innovation, where an innovator searches for a solution or the outcome.

The following steps summarize the basic innovation process to realize breakthrough solutions through innovative thinking:

1. Understand the need for innovation and its purpose.
2. Research a topic individually, collectively or through the networked resources, and gain a deeper understanding of the subject. Do not immediately solve the problem without proper research and knowledge.
3. Identify the potential variables affecting the problem. Make the list as long as possible and expand it using creativity tools, such as benchmarking, brainstorming, mind mapping and TRIZ.
4. Test "what if" scenarios to isolate unlikely combinations of variables and identify likely combinations of variables. The objective is to remove obviously unrelated variables and retain related innovative solutions.
5. Establish the dimension of improvement or the performance characteristic(s).
6. Investigate likely combinations that could improve the performance characteristic(s).
7. Extrapolate the dimensions of interest and validate potential outcomes.

8. Expand your thinking by applying appropriate TRIZ-like principles to explore potential innovative solutions for generating significant change, thus making innovation obvious or disruptive.

9. Continue to explore and formulate alternative solutions. Select a solution that produces expected breakthrough improvement for further validation, optimization and implementation.

The Brinnovation™ approach to developing an innovative solution is more systematic and much faster than searching for a solution from millions of possibilities. The current process of innovation appears to be an art, a random occurrence, or a flash of genius because of the frequency of its occurrence and the unpredictability associated with the search pattern. Some innovators, who have mastered the searching process, look like serial innovators, while those who do not know the process rarely conceive an innovative idea.

The idea generation process in the current environment focuses on ideas about the potential solutions and then picks the one that justifies the use of resources for a tryout or its novelty. The Brinnovation™ process of innovation, however, incorporates a system for creativity or divergence and innovative convergence. The planned convergence process, or an algorithm associated with it, can speed up the innovation process to identify causative sources of innovation. Once the purpose and causes of innovation are identified, then the extrapolation is utilized to achieve the desired extent of innovation.

The current process of imagination focuses more on subtle aspects, such as visualization, dreaming or using the subconscious mind. Actually if a person is introspective, the current process of imagination can be described as an ability to imagine various possibilities. In order to understand the imagination process better, looking at its boundary condition, which is pure imagination, is necessary.

Pure imagination appears to be conceiving very random thoughts or possible solutions and then playing with them in the mind by stretching them to their limits. Typically, when people imagine and stretch, they tend to go to the ultimate limits, which are beyond business needs. People imagining to that extreme often get lost and forget the purpose of innovation as well as their chain of thoughts. Thus, innovation on demand requires purposeful imagination, which is the identification of practical solutions and the extrapolation of the best solution in the direction of innovation.

INNOVATIVE IDEA GENERATION

One of the challenges in developing innovative solutions is the practice of innovative thinking. In many brainstorming sessions, most of the ideas appear to be on the line of "been there, done that, nothing new, and same old, same old." Many ideas or suggestion programs fail because of triviality or the purposelessness of ideas. Many people do not even consider themselves innovative individuals. Even the "perceived dumbest" mind has enough neurons and axons for truly innovative thought. In order to stimulate human thinking, a process, which is simple and powerful but initially perceived to be trivial, has been developed. However, in the many sessions the author has conducted, this simple process works.

The first step is to **clear the mind**. This step is achieved by asking people to write down good ideas they have about a topic without talking to other people. People love this step, as they already have so many good ideas that are clogging their minds. Once these ideas are written down, the mind is open, biases are out, and resistance is down. Reviewing these ideas shows that most of the ideas are "same old" ideas everyone has already thought of and found to be useless.

Having cleared the mind, people are now asked to write crazy ideas about the topic. "Crazy" here is defined as **logically stretching** the mind by thinking about what can be done to the subject of innovation at its extremities. The left hemisphere of the brain usually drives the crazy thinking. These ideas stretch current performance levels. Some people continue the "good idea" process; thus, many of the "crazy" ideas still look like "good" old ideas. Some people really struggle as they try to conceive crazy ideas.

The next step is to involve the right hemisphere of the brain by asking participants to write down stupid ideas. "Stupid" ideas here represent unintelligent ideas, which really are **unrelated to the subject**. Participants see the difficulty in conceiving stupid ideas, and they learn to appreciate stupid ideas (as they really are well-thought-out innovative ideas).

The right hemisphere of the brain usually drives spatial thinking, which broadens the space of innovation. This thinking represents the creative aspects of innovation that people are afraid of thinking about for the fear of being called "stupid." They must recognize that stupid ideas are innovative ideas as well as some of the possible combinations of variables. At this stage, practically everyone tries to avoid looking "stupid;" with enough prodding, however, participants do generate some ideas. The objective is to learn to think innovatively by utilizing all available mental resources and gaining speed of thought by practicing thinking. Some people are good at thinking of stupid ideas. Such individuals have a sense of uniqueness and differentiation.

At this point, people have learned to apply thinking on demand (i.e., flexibility of thinking). Mental agility is fundamental to developing **innovative thinking** quickly. The final step is to write down funny ideas about the subject of interest for innovation. Here practically everyone stumbles, with a few exceptions. People have to think hard to come up with funny ideas. The innovation process looks like an improvisation. At this stage, people are now practicing combinatorial play, or freely trying to associate various things they know about the subject of innovation.

Reviewing several sessions on applying the above process indicated that idea generation does take time. Actually, generating innovative ideas takes even more time than does generating good ideas. Figure 7.8, Innovative Thinking, shows that funny ideas take the longest and are more innovative than are good ideas.

Many leaders like to communicate to their employees that they want them to have fun at work. Measuring whether

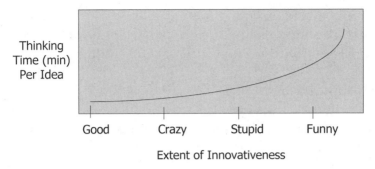

FIGURE 7.8 Innovative Thinking

employees have fun at work or not is difficult. The objective is not just to have fun, but also to have fun productively. Learning to give funny ideas demonstrates what can be a measure of having fun at work. Having fun means employees are free to give funny ideas without any fear. A measure of purposeful funny ideas is a great measure of the innovative thinking of an organization's human capital.

SUMMARY

Henry Chesbrough, in his book *Open Innovation*, talks about the shift in the paradigm of innovation from R&D-driven, closed innovation to open innovation. Accordingly, Open Innovation recognizes value creation through innovation outside the boundaries of an organization. The closed innovation model focuses on mainly internal sources for innovation versus the open innovation model that utilizes the best resources available anywhere. In today's economy, businesses should deploy internal and external resources to speed up innovation. Open innovation incorporates external ideas and internal resources to create value.

Brinnovation™ utilizes the concepts of open innovation and external sources to gain knowledge and ideas, while internal competencies are utilized to create value from the ideas. The process utilizes external resources for enhancing the creative component through networked research and benchmarking and internal resources to develop an innovative solution.

The purpose of Brinnovation™ is to accelerate the frequency of breakthrough solutions in an open innovation environment, while still maximizing the intellectual use of available human capital.

Since Brinnovation™ is based on the experience of the best innovators, such as Einstein, Edison, Galileo, Newton and Ford, it is a benchmarked process. The process is less dependent on an individual flash of genius and more dependent on access to the available knowledge at work. As a result, any individual can become innovative if he or she practices the process of innovation. The framework of innovation identifies the components of innovation and enhances understanding of the innovation process, which eventually will lead to a more robust innovation process. The challenge is in accepting that everyone can be innovative and produce innovative solutions on demand.

TAKE AWAY

1. Our drive to innovate comes out of our fear of extinction, or out of the need to make our life more comfortable and pain-free.
2. Drucker identifies seven sources of innovation that include flash of genius, exploiting incongruity or contradiction, growth in demand, changes in demographics or perceptions, and creating new knowledge.
3. Knowledge innovations require focus, dedication, ingenuity, knowledge, and most importantly, perseverance over a longer innovation time. Over time, knowledge innovations become fragmented and lead to many smaller innovations.
4. In the knowledge age, with access to the Internet, a networked individual is now the building block of innovation.
5. Commercialization of a creative solution makes innovation a random occurrence, while creating a solution to a commercial demand definitely makes the solution innovative. Thus, innovation on demand diminishes the difference between creativity and innovation.
6. Every idea must have some energy associated with it that is an outcome of effort and speed of thought.

7. The speed of thoughts relates to how fast the brain processes the information stored and received, thus generating ideas or possibilities.

8. Based on the extent of activity and outcomes, there are four types of innovation—Fundamental, Platform, Derivative and Variation.

9. Each type of innovation requires a different mix of knowledge, play and imagination. The amount of 'play' required for various innovation activities increases when moving from fundamental to variation innovation.

10. Fundamental innovation takes the longest amount of time, while variation innovation takes the shortest time.

11. Brinnovation is the framework of systemically developing an innovative solution on demand resulting in a change of at least 47.5%.

12. To achieve the breakthrough, one must think beyond 'good' ideas by thinking in crazy, stupid and funny ways.

13. The measure of purposefully-generated funny ideas is a great measure of the innovative thinking of an organization's human capital.

ROOM FOR INNOVATION

LAURIE LAMANTIA

The concept of Innovation Rooms, though not new, is a novel ~~original and of a kind not been before~~ *idea. They are intended to inspire creativity among employees. Organizations like Bell Laboratories, known for their creative talent and leadership in innovative products, have located such rooms in the heart of their facilities. The Innovation Room provides an environment—atmosphere, tools, and stimulation— that helps to unleash the creative talent within individuals to seek better and creative solutions to organizational problems.*

The Innovation Room is a microcosm of the macrocosm. It is a miniature seed culture of the larger organization's desired culture. Whatever is going on in the larger organization ultimately should be no different than what is occurring in the Innovation Room. This room reinforces the innovative intentions of the larger organization, and this reinforcement is very powerful.

Employees use the innovation room as a place to experiment and practice new ways of operating (but in a smaller, *?* more controlled context). Then they return to the organization, which is also transforming, and continue to practice the new things they are learning. As the people change, they help the organization change, and as the organization changes, it helps the people and teams within change.

The Innovation Room starts to breed transformation and helps create momentum for the new corporate way. The people who create the Innovation Room basically ask the question, "Why can't we allow ourselves to tap into the power of the human imagination, creativity, and collaboration?" Thus, they birth the Innovation Room.

Keep in mind, a room for innovation is not the only way to change corporate culture—it is just one way to accomplish this task. But one thing is for sure; the innovation room provides an opportunity to say, "We do things differently around here. We play, we experiment, we share, and we act as one. We are all on the same team, and in here we are learning to act like it. We are learning how to stimulate our imaginations."

CULTIVATING AN INNOVATIVE MINDSET

Innovation Rooms are an excellent way to create innovation mindsets. Inside the Innovation Room, *teaching, inspiring* and *experiencing* desirable innovative actions and behaviors is possible. The Innovation Room provides a variety of experiences and unconventional training, not only expanding the right and left hemispheres of the brain, but also stimulating the mind, body and spirit of the marketer, engineer and customer service representative. The building blocks of an Innovation Room should ideally include the "teach," "inspire" and "experience" components.

TEACH

Teach the behaviors, mindset and tools of innovation.

IdeaVerse, a creativity and innovation center inside of AT&T Lucent Bell Laboratories, provides over 100 different courses that help associates learn and practice the new mindset and tools of innovation. Courses range from a team-creating tool, 6 Thinking Hats, developed by Edward DeBono, to a painting course titled *Life Paint and Passion.*

Foster communication and cross-pollination.

The Chicago branch of LaSalle Bank recognized it was not getting the benefit of people's ideas, so it sponsored an initiative, *The Idea Center,* to gather ideas and act on them. This initiative became a way for LaSalle Chicago to encourage employee creativity by implementing and rewarding ideas submitted to *The Idea Center*. The employee ideas have given birth to many process and product improvements for both the company and its customers. By tapping into the ideas of the people, LaSalle is undergoing a grass-roots transformation that is paving the way for more ideas, more enthusiasm and more good business.

Facilitate team collaboration.

Ideo, a design firm based in California, encourages its teams to brainstorm freely, imagine all kinds of possibilities based on data they gather from real users, and build on one another's ideas in a 3-day product prototyping session the company calls the "deep dive."

INSPIRE

Experiment with new behaviors through risk-taking and change.

Cargill is a private company that has a Chief Innovation Officer. An Innovation Officer is someone whose sole job is to ensure that the company is consciously and continuously striving to be the most innovative in its class. Cargill is so serious about innovation and creativity that it has taken a risk and put an executive in place to foster it.

promote the growth of

Encourage play, fun, laughter and heart.

A company outside of Chicago takes an hour-long dancing break, where the associates stop to line dance in the company cafeteria.

Stimulate imagination.

offer or accept a challenge

Microsoft taps into the creative power of each of its software engineers by throwing down the gauntlet. The company set goals for the engineers that not only challenge them to be *the company* to invent new products, but to be the company that *makes it "cool" and makes it a commercial success*. With these goals, Microsoft is finding a way to tap into the emotional energy of creativity.

Encourage connection-making.

The Ryan Center for Creativity and Innovation at DePaul University has thought-provoking quotes on ceiling tiles and bizarre creations hanging in boxes around the room. Its brightly-colored walls and *quote for the day* wake people up and encourage thinking in new and unexpected ways.

Inspire greatness.

Hewlett-Packard's philosophy for fostering an environment for creativity and innovation includes the following:

> Believe that you can change the world. Know when to work alone and when to work together. Share tools and ideas. Trust your colleagues. No politics and no bureaucracy are allowed. The customer defines a job well done. Radical ideas are not bad ideas. Make a contribution every day. If the idea doesn't contribute, it doesn't leave the garbage. Believe that together we can do anything.[1]

EXPERIENCE

Out-of-box thinking and doing

Enesco's Blue Sky Center for Creativity and Innovation created the *Fun Factory* designed to help teams manufacture creative ideas. The *Fun Factory* used an out-of-the-ordinary-process, like "come up with a new $5 product in 5 days," to push the envelope and see what could happen.

Opening to new approaches and ideas

Steelcase, Inc. designed the building of its Corporate Development Center in the shape of a pyramid. The interior

[1] 1999 HP Annual Report (as captured on www.creativityatwork.com)

was designed to encourage communication and creativity by having both individual office space and team spaces.[2]

SYMBOL OF INNOVATION

The Innovation Room is symbolic on several levels. It is symbolic in that it is there right in the middle of the company, taking up physical space. It is also symbolic in the way it is appointed, decorated and facilitated. The creative space of *IdeaVerse* is a pie-shaped room that has purple walls, beanbag chairs, and furniture on wheels, signaling things are different there. IdeaVerse's creative space contains a library, fresh coffee and a huge white board. This symbolic space helps reinforce what the company is about and what it desires to see happen.

One way to look at innovation is "the fulfillment of a connection-making process that not only creates something novel, but 'pays-off' in some way." The average adult thinks of 3–6 alternatives for any given situation; the average child thinks of 60 alternatives. How do we get adults to make more connections, as making more connections is one of the most important ingredients of breakthrough ideas? The Innovation Room can stimulate imagination and help make connections— lots and lots of unexpected connections—by:

1. Creating a wonderful library of picture books, self-help books and everything in between for people to check out or use on the honor system.
2. Providing an iPod of pre-loaded music for people to sit and enjoy for 5 minutes.
3. Using Feng Sui to encourage chi to flow through the room.
4. Bringing in plants and nature to help people connect to their loves outside of work.
5. Hanging interesting paintings and unexpected art (like a mirror made of bottle caps), ceiling tiles with provocative statements, and hand-painted carpets.

[2] Steelcase website www.stealcase.com

6. Providing childhood toys to help people reconnect with their playful selves. Legos, play dough, tinker toys, etch-a-sketch and others give a sense of the past and help us remember back to carefree days. Toys and mind-challenging games all give the brain a break and stimulate fresh ideas.

7. Using the Internet and networks to check out blogs, interactive websites, and other unexpected ideas people have created on the web. Check out www.fusionanomaly.net or www.wikipedia.org or www.slashdot.org.

ENHANCING CREATIVITY

In the Innovation Room, remembering the power we have to create is possible. One of the most vital and underutilized tools is our creativity. Would it not be great if people really understood their creative power? Would it not be great if they applied creative tools to the business problem? The Innovation Room helps empower and remind people of their innate, natural power . . . their power to create, their power to make a difference, their power to change things for the better. The offerings and experiences of the Innovation Room provide people with the following:

KNOWLEDGE

Classes help people think creatively and learn the building blocks of the creative process. People are taught about the brain and how it works. They also learn about their creative preferences and the contexts in which they feel most innovative. Knowledge is power, and the Innovation Room provides people with new knowledge and helps them unlock their internal power.

So much talent and possibility is locked inside of every person and every organization. Much of what the Innovation Room provides is an opportunity for people to slow down and connect to this wisdom deep within—to connect to their unique ways of contributing.

Play

The Innovation Room provides an opportunity to play and remember one's creative self. Often we remember times when we were kids. We painted and constructed Lego forts. Encourage people to reconnect to that playful creator inside themselves.

Imagination

The Innovation Room helps spark the imagination of people through the various connection-making tools and items. Stimulating the imagination is really quite easy and quite difficult at the same time. The easy part is that the brain loves to make connections and is exceptional at it, because it is constantly looking for patterns and recurring themes. Any random thing can be used by the brain for connections *if* we give ourselves permission to make the connections and time to free-wheel. Most people, however, will not give themselves permission to play with possibilities.

FOSTERING COLLABORATION

One thing that most teams need, but do not seem to have the time or resources to get, is help—help facilitating them through difficult issues, help moving them through a problem-solving process, and help finding and teaching tools for working more effectively together. The Innovation Room provides facilitators who can facilitate teams through various processes that will help them move forward.

A short list of processes that benefit teams greatly are:

- **Creative Problem Solving**—This 6-step problem solving process developed by Alex Osborn is taught through the Creative Problem Solving Institute. www.creativeeducation-foundation.org

- **Scenario Planning**—This is a process, developed by the Global Business Network, in which different scenarios are developed as tools for ordering and planning for alternative future possibilities. www.gbn.org

- **6 Thinking Hats**—As delineated in Chapter 3, this process separates thinking into six categories for analyzing issues and generating new ideas. www.debonoonline.com

- **TRIZ**—This is a set of tools for developing and evolving systems, solving problems and selecting solutions. www.triz.org

The key is to find *communication and collaboration processes* that keep teams growing and enlivened. Conscientiously learning and using good communication processes gives structure and effectiveness to team interactions. Look for offerings that nurture respect, deeper listening, communication, and a sense of community among team members. Collaborative processes that teach problem solving, idea assessment, decision making, meeting effectiveness, and other team-governing skills are excellent investments.

FACILITATING INNOVATION

The Innovation Room can encourage and help people practice these new behaviors in the safety of a supportive environment. The Innovation Room and its mentors can help to breed these new behaviors; however, if the overall company environment is not also in support of these new ways, the seeds will fall on fallow ground. One of the most frustrating things for people is to go to a training or wonderful learning experience only to return to their companies, have people ridicule them, and make them feel like their new ideas are not welcome.

People get frustrated when learning new ways to approach problems and opportunities if the rest of the organization continues to operate in a "business as usual" manner. "Why bother" becomes the defeated attitude. Learning something that can really make a positive difference, only to have it rejected by the organization, is the worst possible outcome.

The Innovation Room should not face the constant issue of "Prove this is a valuable use of company resources." This "prove it" mentality drains and wastes valuable company energy. An incessant need for proof takes the heart and soul right out of something that is important and necessary. Any place where people can get together, play, have fun, learn, grow and collaborate has to be good for business.

The Innovation Room can measure new ways of doing business through feedback mechanisms, looking for such things as:

- Learning new tools or processes that will help in day-to-day work
- Experiencing collaboration
- Experiencing play and rejuvenation
- Experiencing an "aha" moment or breakthrough concept
- Making a new connection for a current project
- Learning tool(s) for better communication
- Feeling more creative and able to create

DEVELOPING DIVERSITY OF THOUGHTS

The Innovation Room and its activities, classes and offerings need to speak to the whole person. We are not just engineers or marketers; "we are huge and we contain multitudes"[3] beyond our wildest imaginings. The opportunity to speak to the whole person is an opportunity to stimulate greatness.

The Innovation Room *welcomes diversity and encourages whole-brain thinking* by providing training to showcase different thinking and "doing" styles.

- *Herrmann Brain Dominance Indicator* (HBDI), a great tool for self-discovery, is a personal assessment to understand thinking and creating preferences.[4]

[3] Walt Whitman
[4] www.hbdi.com

- *The Myers-Briggs Type Indicator®* (*MBTI®*) is a forced-choice personality inventory based on C. G. Jung's theory of Psychological Types. Its purpose is to make this comprehensive theory of personality practical and useful in people's lives and indicates a personality preference as one of 16 Types.[5]
- *The Enneagram* delineates nine basic personality types of human nature and their complex interrelationships. The Enneagram is also a symbol that maps out the ways in which the nine types are related to each other. This tool provides a framework for understanding oneself and others.[6]
- *The Diversity Game* is an entertaining, informative game designed to present a picture of a group's thinking style preferences and helps point the way towards working together more effectively and productively.[7]

Keeping an "open door" policy for the Innovation Room is important to everyone. Keep sending the message that the Innovation Room is a resource like the public library—available to everyone, utilized by everyone and benefiting everyone. Humans like to categorize and segregate, which can happen with something as simple as an Innovation Room. People start to put a box around the room, turning it into the exact thing it was meant to eliminate—limitation and segregation. Look out for assumptions people make that the room is . . .

1. Only for "idea people"—where the creative go to be more creative.
2. Separate from the rest of the way the company does business.
3. For the "elite" or "special" projects.
4. Where you "go to get ideas," or the only place in the company where idea generation happens.
5. Only for people in marketing or R&D, because they are the creative ones.

[5] Peter Geyer, INTP
[6] www.enneagraminstitute.com
[7] www.hbdi.com/diversity.html

One of the most helpful things the Innovation Room can provide is helping teams connect to what they want to accomplish and how they are going to accomplish it. Envisioning together, in a retreat of sorts, helps teams get out of their daily routines and see things in a new light. The Innovation Room offers visioning sessions for teams and individuals to help them connect to their meaningful and important work. The Innovation Room can also provide experiences and mini-retreats for individual associates to explore what they want to contribute to the organization's mission.

In any creative process, a time comes when uncertainty occurs. Working with the uncertainty, welcoming it, and then using it to highlight new possibilities (a new product, process, service or thought) is a muscle the Innovation Room helps to exercise and strengthen. In *Creative Problem Solving* sessions or *Team Visioning Sessions*, teams reconnect to their purpose and vision and let the uncertainty combine with the problem to create opportunity and breakthroughs.

MAKING TIME FOR INNOVATION

An important part of the creative process is incubation. Whenever someone is asked, "Where do you get your best ideas," people usually answer something like "in the car, in the shower, or while running/walking/exercising." In the "down time," the brain takes all the data it has and starts to make connections. *We all need time to let the creative connections happen.* The Innovation Room provides this opportunity. People can come and just relax, hang-out, read, listen to a new video, and do the equivalent of stew or marinate—letting all that data and input fuse together to create something fresh.

At Maddock/Douglas, a new product innovation firm in Elmhurst, IL, little incubation rooms are peppered throughout the facility. The rooms have comfortable, colorful chairs and furniture one can write on to focus and noodle with ideas in private.

Time in the Innovation Room needs to be about fun as much as it is about learning, growth, and transformation. Be open to making changes when things do not seem so dire. In these rooms, we can let down our guard and begin to practice the things we are learning—playing with them and seeing how they work versus having to get it right all the time. Innovation Rooms can contain ping-pong tables, dartboards and pool tables. A company in Holland has kept the top floor of its corporate offices for the people and not the executives, putting in a baby grand piano for company gatherings.

PRACTICING CHANGE

The Innovation Room represents opportunity, a chance for change and transformation, and the ability to take the old ways and make them new. The Innovation Rooms are unique in and of themselves, because not only do they spark new ideas, products and innovations, but they also are an innovation, sparking a new way of doing business. We are doing things we have not done before. We are questioning unquestioned assumptions and asking if the old ways work anymore. What are more supportive ideas that bring us where we say we really want to go?

In the Innovation Room, people operate in a new way and learn things that traditional corporate America might find uncomfortable (at best). People do business in a new way, listen to their passions, respect the ability to choose, and band together as a team. Many things in the Innovation Room are nontraditional—even radical. Resist negative labels. Operating with such intentions will seem radical to the old guard—not too far in the future, however, these intentions will be standard operating procedure (SOP) and considered beneficial.

The disciplines necessary to create a new kind of corporate environment are very different than the disciplines we have practiced to this point. We need to summon courage and persist in the face of what might seem to be foolish ideals and unrealistic hopes. An innovative environment is possible to create, but the brave need to decide to go first and usher innovation in. They are paving the way for the rest. The Innovation Rooms—what they symbolize, what they teach, and what people come to learn and practice in them—will foster a new corporate climate that is innovative, creative and prosperous.

TAKE AWAY

1. Since the innovation process is becoming routine, a designated space, so-called the "Innovation Room," will be needed to facilitate innovation.

2. The Innovation Room reinforces the innovative intentions of the larger organization, and this reinforcement is very powerful.

3. Innovation Rooms can be used to *teach, inspire* and *experience* the kind of innovative actions and behaviors one would like to see people incorporating into their daily skill sets.

4. The creative space of *IdeaVerse at AT&T* was a pie-shaped room that had purple walls, beanbag chairs, and furniture on wheels that signaled things are different there. Also included were a library; fresh coffee and a huge white board.

5. The Innovation Room can stimulate imagination and make connections with possibilities.

6. The Innovation Room empowers and reminds people of their innate, natural power . . . their power to create, their power to make a difference, their power to change things for the better.

7. Innovation Rooms provide an opportunity for people to reflect on and connect with this wisdom deep within—to connect to their unique ways of contributing.

8. The Innovation Room can encourage and help people practice their new behaviors in the safety of a supportive environment.

9. The Innovation Room *welcomes diversity and encourages whole-brained thinking* by providing training to showcase different thinking and doing styles.

10. Keeping the room open to everyone and continuously sending the message that the Innovation Room is a resource like the public library—available to everyone to come, utilize and benefit from—are important features.

11. The Innovation Rooms help people practice feeling those times when "we don't know," thus working through that uncomfortable time to where we can make the most of that discomfort.

12. Time in the Innovation Rooms needs to be about fun as much as it is about learning, growth, and transformation.

13. It is in these rooms that we can let down our guard and begin to practice the things we are learning—playing with them and seeing how they work.

14. Opportunity is represented in the Innovation Room—a chance for change and transformation as well as the ability to take the old ways and make them new. The Innovation Rooms are innovative in themselves.

15. Bravery is involved in having Innovation Rooms, in that we are doing things we have not done before. We are questioning unquestioned assumptions and asking if the old ways work for us anymore.

INNOVATION DEPLOYMENT

Praveen Gupta

Most corporate leaders recognize the need for innovation to create a sustainable competitive advantage. Typically this responsibility is delegated to just the R&D department. Research, however, shows that every member of the firm has the capability to contribute to the firm's success and be innovative in how he or she approaches his or her function. Thus responsibility falls on the leaders to provide the environment and the opportunity to capitalize on this innovation capability. The individual results may vary from incremental to breakthrough, but they will all help the firm become competitively stronger. The leaders need to build an organizational framework that will encourage innovation. This effort will require taking the initiative to promote an innovative culture, create a supportive organization structure, institutionalize a process, provide the requisite training to the employees, measure the results and reward the contribution. People truly become the "most valuable asset" when most employees feel they are being innovative and contributing to the firm.

INTRODUCTION

Innovation is a necessity in maintaining a competitive edge. Customers are becoming more demanding and restless. Customers want unique products when they need them. In

183

response to this demand, companies like Nokia, Sony, IBM, Apple, 3M, and Proctor & Gamble have been innovating for a long time. Some companies innovate by allowing 15% time for independent projects, they buy innovation, or they collaborate for innovation. Historically most innovation is a function of research and development (R&D) departments. Businesses recognize today that innovation must be institutionalized.

According to a survey conducted by Chasan (2006), corporate leadership struggles to raise the profile of innovation in their companies due to internal barriers, such as culture and climate. Yet corporate leaders are looking to innovation in their business models to drive growth. About one in seven of the participating CEOs think internal R&D is a good source for innovation. Moreover, many CEOs are unwilling to make innovation their top priority because of the lack of understanding of the innovation process, and only one in five in the United States and India want to take ownership of the innovation. These leaders have seen that the most-currently-used measures of innovation do not strongly correlate with the top or the bottom line.

In an international symposium on innovation methodologies (Eurescom, 2002), the main challenges appeared to be the lack of knowledge in innovation techniques and the sharing of innovation results. To overcome these challenges, the suggestion to establish a "European Innovation Award" and create an innovation toolbox for accessing innovation knowledge was forwarded.

A review of literature and other various sources shows that as many innovation methodologies exist as the number of users or organizations. Since many people consider innovation an art, every innovator or innovating organization develops a unique approach to innovation. Most approaches appear to have common elements; however, the details of the process and an understanding of its components are lacking. Currently, a strong correlation between types of methodologies and their performance is not established. The question then remains: Which innovation approach is a good methodology? What should the criteria be to evaluate a methodology of innovation?

In order to evaluate a methodology, it must support some theoretical objectives, some basic tenets must form its foundation, and the establishment of relevant measures must be present to monitor and improve the innovation process for improving its efficiency and its outcome. In other words, a repeatable and continuously improving system must be in place, allowing standardization of structure and discipline for the innovation process. Such a system must support the teaching of it to others, so that everyone clearly understands the innovation methodology.

The breakthrough innovation methodology, called Brinnovation™, was formulated based on its implementation at various businesses and the logical understanding of its steps. As explained in Chapter 7, Brinnovation™ bases its tenets on the following assumptions:

1. Innovation and creativity are the same when produced in real time on demand;
2. Innovative solutions can be generated on demand in a knowledge economy;
3. The building block of innovation is a "networked individual;" and
4. Innovation is a function of speed of thought.

Every individual or group of individuals is fundamentally a creative entity due to the colossal capability of the brain. No two tasks accomplished by any one person are identical. Developing a creative solution on demand is an innovation, as it is bound to create value. Instead of developing a solution and then commercializing it, innovation on demand implies identifying an opportunity first and then inventing a solution for the problem. A group of individuals thinking independently and working together is more effective than a group of people brainstorming (thinking together) and working independently. Therefore, a cluster of people by itself does not become more innovative; instead, a network of thinking individuals is more innovative. Such a network enables individuals to learn quickly, think independently, collaborate virtually, and innovate when needed.

For individuals to become active thinkers, they need to have broader experience and must be able to practice combinatorial thought experiments. Since the possible combinatorial thought experiments can be numerous, an active thinker needs to think fast. The speed of thought matters and is affected by knowledge, which allows us to develop shortcuts to prioritizing information and conducting thought experiments faster.

Thinking and conducting thought experiments take time. It requires motivation, stamina and a rested mind. With this understanding, corporations must create an environment where people can think freely. One of the main inhibitors to thinking freely is fear, which executives must try to drive out of their organizations. Driving out fear, however, does not mean removing accountability for actions or consequences.

Many businesses begin when a person with an innovative idea takes action to realize it. Diversification of the business helps promote business growth. Businesses experience many up-and-down performance cycles, including cycles of loss and profit, growth or downsizing, and mergers or acquisitions. Through these cycles, businesses go through many changes, which can lead them to profitable growth or a hard landing. The hard landing can be due to the management style, unexpected market erosion, loss of a major customer, externally funded growth, or simply a lot of waste because of poor management. Profitable growth, on the other hand, requires innovative strategies and solutions.

The earlier model of R&D driven product or service development cannot keep up with market demand for innovative solutions. Global connectivity, knowledge sharing, trade and opportunities change the former business model; now the current business model involves global customers, global development and global innovation. Innovative thinking and activity-based on innovation must become a routine. The marketplace now expects employees to produce innovative output on demand, with value and on time. The time of innovation can be from real time to years on demand. The innovation can be a simple variation to a fundamental discovery of a new phenomenon.

In today's knowledge economy, business orientation is shifting from manufacturing to service. Businesses, whether manufacturing or service, gear themselves towards managing intangibles more than managing widgets. More importantly, the intangibles include knowledge management and innovation. In a conventional business model, innovation is a function within the R&D community. The role of innovation has changed in today's knowledge economy. Experience shows that now everyone's job is to innovate at relevant levels in an organization. In today's environment, innovation is a tool for creating the unique selling proposition (USP). Innovation is driving the competitive bar to higher levels.

ENCOURAGING INNOVATION IN THE ORGANIZATION

To achieve higher performance continually, business leadership must be cognizant of the intellectual potential of employees. Committing to innovation throughout the organization will accelerate the performance, as more employees are collaborating with the leadership team rather than resisting it. Such interaction in the organization reduces friction among employees and managers, identifies new opportunities for innovation, reduces cost, and creates a sense of ownership among employees.

Commitment to continual innovation requires a good understanding of theory, practice and the results of the innovation methods. Scattered and successful application of innovation methods demonstrates that the innovation process is more than random creativity. Just like any other process, one can benchmark the best innovators and organizations to learn and innovate. Einstein happened to be the best thinker, and Edison was the best innovator. Einstein's discoveries are more fundamental, but Edison's work was product-based innovation.

How did Edison innovate so frequently? He understood the innovation process, built his laboratory in Menlo Park, NJ, and guided his researchers to produce solutions on demand. He built factories and products based on his innovations and accelerated that growth in wartime. Edison was not acting as an innovative person; instead, he was seeking opportunities and producing innovative solutions. He expanded his knowledge, changed his expertise from one field to the other, and innovated on demand.

To institutionalize innovation on demand, benchmarking against Einstein and Edison is important. Einstein mastered thought experiments and saw something in nothing, and

Edison perfected product innovation to produce innovative solutions on demand, as evidenced by the number of patents (over one thousand) assigned to him. The corporate process for breakthrough innovation must consider the following aspects:

DEFINING INNOVATION

Clayton M. Christensen and Michael R. Raynor, in their book, *The Innovator's Solution*, emphasize sustained innovation in achieving corporate business growth (Christensen, 2003). A successful era of superior performance in the life of a corporation occurs due to some innovative disruption. Sustaining innovation requires not just the ideas, but also the packaging of ideas for growth opportunities. Even Six Sigma emphasizes breakthrough improvement; however, methods do not currently exist to produce breakthrough solutions.

Innovation is often defined as open-ended. Everyone has a different perception of innovation and creativity. The extent of innovation can be incremental (implying a little change), radical (representing a disruptive change), or of a general purpose (implying a new discovery). One way to define innovation is "doing differently." The question then is how much change should the innovation create? Considering the process of evaluating change, a change is statistically significant when the change exceeds at least 47.5% in the desired characteristics. The 47.5% change corresponds to two standard deviations from the current process' typical performance, as shown in Figure 9.1. When the change in a parameter is statistically significant, the probability of occurrence is small, and the change is a breakthrough innovation. Thus, innovation happens when an activity occurs differently in order to create value through products, services or solutions.

Innovation happens daily, whether it is a simple container by Rubbermaid, an Apple iPod, Panasonic's robust laptop, a cell phone from Motorola, Sony, LG or Nokia, a grill from Weber, new drugs from a pharmaceutical company, or a printer from HP, Canon or Xerox. Many companies are serially innovative, while others are sporadically innovative. Thus, innovation is not a new thing, but it is often sporadic. In order to acceler-

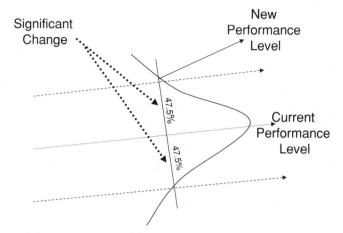

FIGURE 9.1 Breakthrough Solutions

ate and sustain innovation for meeting continual demand for new products, services or solutions, corporations must deliberately recognize organizational needs and address them. This recognition requires an understanding of strengths and weaknesses, company leadership for innovation, and a clear link between innovation and the corporate values and organizational strategy.

In order to understand an organization's strengths and weaknesses, an assessment must occur using simple tools like a checklist, survey or performance analysis. A thorough analysis must include social, operational, financial, customer, and leadership aspects of an organization. The social aspects may include corporate values, teamwork and employee participation. The operational assessment includes an emphasis on creativity in process management and daily activities, the ability to take risks, and a general decision-making approach. The financial aspects include resources committed to innovation-related activities, training, rewards, and revenue generated from innovative solutions. The leadership aspects include creativity at the leadership level, inclination for risks, recognition for success, understanding for genuine innovation failures, and keen participation in innovation. The assessment's objective is to understand how an organization can transform into an agile and thinking organization. A thinking organization is one that promotes learning new skills, experiencing new domains and productively applying lessons learned for developing innovative solutions.

In a training session, people are always looking for a trick, a case study, software, or a formula to apply quickly so they can reject the tool for its differences and application difficulty. Quickly applied methods, however, are not adequately developed and will not produce desired results. This rote application of a technique is called reproductive thinking. Accordingly, if everyone learns a technique to design a product and applies it the same way, then the expected result will also be predictably the same.

Every problem and every company are different. Thus, the application of a technique must be adapted creatively to the opportunity in consideration. Figure 9.2 illustrates different thinking types. A problem can be solved just by doing something, or many impractical creative ideas can help to solve the problem. The case of "just doing something" to solve a problem often leads to new problems, though, while in the case of using "imaginative ideas," nothing really happens. Thus, the solution needed for the problem often lies in the application of creative ideas.

One of the challenges in a corporation is to allow time for thinking. Companies often hire smart people, keep them busy in fighting fires, and give them no time to think. The 3M

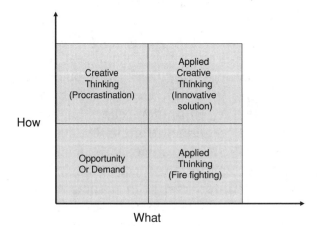

FIGURE 9.2 Thinking Types

Company allows employees to spend 15% of their work time as they wish thinking of something new, learning something new, or doing whatever they like. In order to justify time for thinking, or the investment in innovation, following the approach of systems thinking, as shown in Figure 9.3, must take place.

As the figure shows, innovating a solution for an opportunity at a level higher than the level of its symptoms is essential. For example, if a department is having a problem at one network node, the innovative solution should be implemented above the node level (i.e., at the server), so all nodes can benefit from it. Similarly, if an opportunity is identified for an innovative solution, the opportunity must be defined at a higher level. This elevation of the opportunity creates more value, due to its expanded scope of application, and justifies resources for developing the innovative solution. Once the solution is developed, then it is applied to specific situations.

Systems thinking requires that the corporate leadership practices process thinking, establishes measurements for monitoring performance, and promotes risk-taking in developing innovative solutions. In today's economy, diffusion of opportunities globally mandates that every society continually innovate in its domain of expertise in order to create value. Otherwise, by the law of diffusion, migration of opportunities from higher-cost to lower-cost locations will cause societal frustration.

To maintain market leadership, a company must launch a multi-pronged approach to grow the top line as well as the bottom line. Many improvement efforts to perfect the bottom line

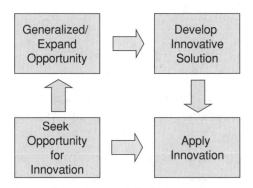

FIGURE 9.3 Systems Thinking

eventually lead to a business downsizing due to lack of sales. Businesses must develop new products, services and solutions. The demand for innovative products or solutions has become a norm. To commit to continual innovation requires a good understanding of theory, practice and the results of the innovation methods. Scattered and successful application of innovation methods demonstrates that the innovation process is more than random creativity.

An organization attempting to institutionalize innovation must determine its methodology of choice. As many innovation methodologies are out there as there are innovation consultants. Therefore, the reasons for selection must be based on a better understanding of the innovation process. Some methodologies are measurement heavy, which leads to a number-driven innovation approach without a predictable outcome. A successful innovation methodology must incorporate inspiration from leadership, involvement of employees, and outcomes for higher value. Such methodology will include planning, organization, a process, tools, measurements, collaboration and celebration.

BUILDING AN INNOVATIVE ORGANIZATION

Planning

How to implement innovation processes for producing innovation on demand is somewhat unknown. Successful implementation requires understanding the innovation process well enough to teach others and multiply resources. The list of challenges in institutionalizing innovation can be exhaustive. Some of the common issues are the following:

- Too much focus on the bottom line and cost reduction,
- Wrong measurements of good performance; for example, headcount reduction is a measure of lean implementation (an improvement methodology),
- Lack of focus on revenue growth through innovative products, services and solutions,

- Lack of strategic intent to institutionalize innovation,
- Inadequate understanding of the innovation process,
- Fear of failure and punishment,
- No time and expectation for the intellectual involvement of employees,
- Insufficient incentives and rewards directly linked to innovation,
- Poorly performing systems used to test new things effectively.

Considering the issues associated with the lack of innovative practices in corporations, a different approach must be taken to plan for institutionalizing innovation. With a clear commitment to innovate new processes, products or solutions, the corporate leadership must develop a strategic plan. In order to approach innovation as a scientific process, a strategic plan must be developed to address the following:

- Strategic commitment
- Organizational alignment
- Measures of innovation
- Plan for innovation
    - Culture of creativity
    - Innovation room (InnoRoom™)
    - Establish innovation friendly policies
    - Communicate innovation
    - Incentives for innovation
    - Demand for innovation
    - Brinnovation™ training and certification
    - Excellence in the idea management process
    - Innovation management
    - Rapid commercialization of innovation
- Return on Investment (ROI) in innovation management
- Strategic adjustment

Leadership for Innovation

Successful leaders recognize the significance of innovation and the required leadership needed. The leader must believe in and understand the role an innovative culture can play in the growth of a corporation in future years. Such leaders consider innovation in all areas of business in creating a culture of innovation. To lead an organization towards becoming a learning and innovating entity, the organizational environment must influence thoughts, planning and acts. For example, Johnson Controls, an organization that has lasted over a century, recognizes the role of innovation (as stated in its values) as responding to its customer needs through improvement and innovation.

To launch or sustain the innovation initiative, the leader must commit to recognizing the intellectual involvement of all employees, valuing all information available, and appreciating the evolution of all employees and processes. The leader of an organization sets beliefs, initiatives, and the environment for innovation. A visionary leadership develops both a corporate meaning of innovation in the organizational context as well as a corporate strategy for learning and innovating success. The leadership establishes expectations and recognition for innovation from employees at all levels. The strategy involves training, recognition, innovation expectations and objectives, the roles of executives, managers and employees, intellectual property aspects, and the transformation from innovation to commercialization of product or service for realizing economic benefits. Executives and managers can set an example through their own behaviors, attitude, innovative thinking, actions, and support of innovation.

Organizational Structure for Innovation

In order to integrate and promote innovation into normal daily activities, the organization must create an innovation model with resources allocated to make it work. In other words, the innovation must become an element of business profitability and growth streams. Innovation begins with ideas, so a mechanism to generate ideas from all employees of the company must be in place. Reviewing the ideas for their relevance and applicability is the next step, and each idea must receive supportive feedback. Employees must be encouraged to think and reflect on their experiences and look to the future with new ideas and innovative products, services or solutions. Innovative or learning organizations support some kind of in-house library, where employees can browse through learning resources to reenergize themselves intellectually.

A recommended organizational structure can incorporate the elements shown in Figure 9.4, which illustrate that the culture of creativity cultivates ideas—ideas that become innovations and innovations that transform into products or services for economic gain. Though developing such an organizational structure only makes sense, many organizations lack this formal structure for establishing a culture of creativity, idea management and innovation leadership.

Culture for Innovation

The culture for innovation begins with an awareness of organizational commitment to innovation and a visible room for innovation as shown in Figure 9.5 (and discussed in Chapter 8). A typical house layout includes a variety of rooms, such as a study room, a kitchen, bedrooms, bathrooms, a living room,

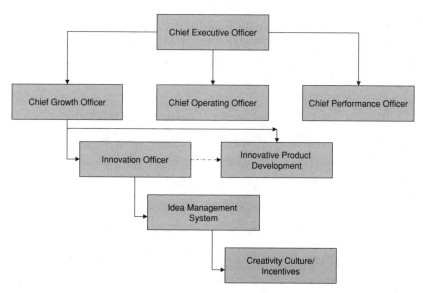

FIGURE 9.4 Organization for Innovation

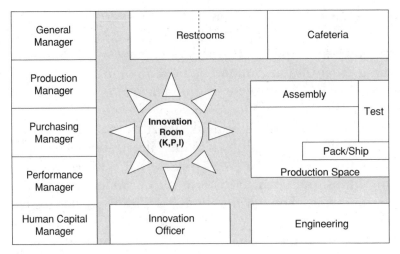

FIGURE 9.5 Room for Innovation

a family room, and a den for relevant routine activities. A kitchen gives people ideas about food. A study room stimulates ideas about the topic of study in mind. Similarly, if an organization likes to have ideas from its employees, it needs to create an environment, or even a room, for creative ideas.

The environment must be one that promotes a sensory experience and knowledge in the field of interest. The leadership desires employees to develop observation-fueled insights, a keen eye for details, and inspiration to do things differently. The objective is to innovate on demand with purposeful effort. According to the analysis of the innovation process and its impact on the brain in Chapter 6, the following conditions must be met for an organization to become innovative:

1. Comfortable environment for absorbing information
2. Effective incentives for learning and comprehension
3. Healthy employees and time management
4. Knowledge expansion and experimentation
5. Resting environment and time for reflection
6. Corporate values and decision making

Innovation requires work, people are innovative in their domain of expertise, and innovation does create economic or social value. Innovation occurs from determined, focused and deliberate work demanding persistence, diligence and commitment.

Process for Innovation

Business is a collection of processes, including the innovation process. Therefore, applying the principles of process management (4Ps—Prepare, Perform, Perfect, and Progress), as shown in Figure 9.6, to the innovation process helps to understand the components of innovation. Like any other process, the innovation process requires inputs in terms of tools, information, material, methods, and skilled people. Creatively deciding upon these inputs is important.

Loosely defining the method of innovation allows for maximum human creativity. The flexibility to learn, experiment, fail and innovate within some defined framework is essential. The innovation process must include experiencing a variety of things outside the norm or domain of work, creating combinations or associations, and validating outcomes. The mental message of various concepts or models results in some practical ideas that can help in formulating products, processes or services. Ultimately, every employee has the responsibility to play and create value through innovation.

If the innovation turns out to be impractical or too costly for further implementation, the originator of that innovation

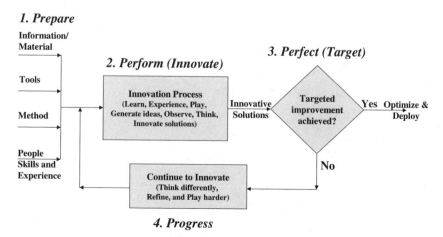

FIGURE 9.6 Process for Innovation

must not become dejected or disappointed. The creative or innovation process itself must be a lot of fun (rather than waiting to have fun with the resulting product). Generating one truly innovative product or service requires many ideas. Therefore, creative play is a necessity, idea generation is imperative, and engaging in innovation is every employee's overriding responsibility to his- or herself, the organization and society. Even if ideas do not turn into products or services, people need to continue to play or create. The persistence and perseverance will eventually lead to innovation on demand.

Measuring Innovation

Because innovation is a loosely defined process, it is a difficult to measure. Innovation is often measured in terms of ideas generated, patents filed, engineering awards given, new products introduced, revenue from new products earned, number of people deployed in innovation, or hours allocated for innovation. To be effective, the measurements must be established as needed in any organization. The intent of measurements must be to assess the role of innovation in the growth of a corporation. The Six Sigma Business Scorecard (Gupta, 2003) identifies a set of measurements relating to growth and profitability. Innovation is a critical aspect of the Six Sigma Business Scorecard. The ten measurements utilized in determining the Business Performance Index (BPIn) are as follows:

1. Employee recognition by CEO
2. Profitability
3. Rate of improvement
4. Employee recommendations
5. Purchase ($)/Sales ($) ratio
6. Suppliers' quality
7. Operational Sigma
8. Timeliness
9. New business/Sales ratio
10. Customer satisfaction

Measurements 1, 4, and 9 relate to innovation in an organization. The CEO recognizes an employee based on innovative solutions with a significant and visible impact on corporate performance. Employee recommendations measure the intellectual engagement of employees, and new business measures the financial outcome of the innovative product or service. A combination of three measurements can create an initial innovation index that has significantly positive impact on the corporate performance.

Training Employees in Innovation

Any major corporate transformation begins with education to ensure learning, consistency, productivity and results. The objective of training is to help people become familiar with the innovation process, accentuate their capabilities, and direct their creativity in the direction of corporate goals. Innovation training should include an understanding of the aspects of innovation in order to become familiar with its components (i.e., knowledge, play and imagination).

The training for innovation should include hands-on experience in researching, playing with combinations of components of the innovative solution, and being in an environment for imagining new solutions. The training may simply be getting people in a planned environment, giving them the learning objectives, and turning them loose. The innovation training certainly must be a creative approach in itself. Irrespective of the method of innovation training, the corporation must set goals for innovation training and measure the training effectiveness in terms of the number of innovations, the magnitude of innovation, and the financial impact.

Recognition and Rewards

Sustaining innovation initiatives depends on the continual success of activities, excitement for doing the innovative work, and the involvement of everyone. As expectations for innovation in various functions are established, the leadership must verify performance against the expectations and make appropriate adjustments. If an organization commits to innovation as a key component of its business strategy, it must ensure that various phases of the innovation process are executed with excellence.

In order to promote creativity or innovation, observing innovation successes is critical. Publicizing success is as equally important as understanding the failures. In a corporation, when creativity, innovation and risk-taking become basic principles, established measures must exist to recognize and reward innovators. Recognition can be as simple as a "thank you" note, some public recognition in a banquet, or an announcement in the local newspaper. Each success is recognized differently—sometimes with financial incentives and other times with personal notes.

Irrespective of the value or type of recognition, recognizing a specific act or outcome of creativity or innovation is essential. The act or outcome can be at the idea level, solution level, or the field performance level. Incentives can be given for submitting an idea about process or product improvement, writing and publishing a paper in a magazine, obtaining a patent, successfully completing the evaluation of a new product or concept, providing new ideas about daily activities, engaging in a superior act of engineering, or transforming an innovation into a commercial product or service. Ultimately, innovation must be an empowering, rewarding and enriching experience for everyone involved.

LAUNCHING AN INNOVATION INITIATIVE

Innovation begins with the intellectual involvement of employees through their ideas. The process of getting employee suggestions, ideas or recommendations has been in existence for a long time. However, its effective implementation and success have been far from satisfactory for many reasons, including the lack of understanding of its value and importance for improving corporate growth and profitability, and the lack of an established process of idea management.

Just like purchasing, sales, production or quality processes, innovation must become a standard process in a corporation. Leadership commitment must exist in order to implement the innovation process. Then an innovation policy must be defined; expectations must be established; resources should be allocated; and measurements must be established to monitor innovation value. Most importantly, incorporating innovation into the business planning and budgeting in order for it to become visible on the management radar is critical.

The first step in creating innovative thinking in an organization is to establish a good idea management program. A high-quality idea management program creates a long-lasting positive impression on employees through sincerity, follow-up and recognition. The purpose, scope, responsibility, ownership, tools and procedures for the idea management process must also be established. In this way, the handling of unacceptable or not-so-good ideas can be clearly defined and documented, and the conversion of good ideas into viable economic value is realistic through training, communication and other business processes. An idea program is not about complaints, criticism of management, or getting even with workplace enemies. Idea management is all about contributing toward employees' success by achieving improvement.

Every employee in a corporation is capable of being innovative. Everyone, at one time or another, has experienced pride regarding something he or she has accomplished. Bringing out the ability to achieve significant improvement requires an expectation that leadership must establish and strive to achieve. The intellectual participation of employees must be a leadership mantra. All successful leaders see potential in their employees and exploit it as the only way to achieve sustained improvement.

TAKE AWAY

1. Businesses recognize today that a new process for innovation must be learned in order to practice and produce more innovation.
2. Many CEOs are unwilling to make innovation their top priority, as currently-used measures of innovations do not strongly correlate with the top or the bottom line.

3. Committing to innovation throughout the organization will accelerate the performance, as more employees are collaborating with the leadership team rather than resisting it.

4. Edison understood the innovation process and guided his researchers to produce solutions on demand.

5. Einstein mastered thought experiments and saw something in nothing, and Edison perfected product innovation to produce innovative solutions on demand.

6. One way to define innovation is "doing differently." Considering the process of evaluating change, one can define innovation statistically when the change exceeds at least 47.5% in the desired characteristics.

7. One of the challenges in a corporation is to allow time for thinking. Companies hire smart people, keep them busy in fighting fires, but give them no time to think.

8. To approach innovation as a scientific process, a strategic plan must be developed.

9. Employees must be encouraged to think and reflect on their experience and look into the future for new ideas and innovative products, services or solutions.

10. A group of individuals thinking independently and working together is more effective than a group of people brainstorming (thinking together) and working independently.

11. Irrespective of the method of innovation training, the corporation must set goals for innovation training and measure the training effectiveness, which should be measured in terms of the number of innovations, the magnitude of innovation, and the financial impact.

12. Just like purchasing, sales, production, or quality processes, innovation should become a standard process in a corporation.

13. Innovation must be incorporated into the business planning and budgeting to become visible on the management radar.

MEASURES OF INNOVATION

Praveen Gupta

The corporate scorecard for most firms does not include measures for innovation. Effectively measuring innovation is difficult, but attempts have been made. Historical measures have included the following—percent of revenue spent on R&D, number of patents, number of R&D initiatives, and number of new products—but they are potentially flawed and do not truly measure the value of innovation. Effective measures of innovation are based on an understanding of the innovation process (unique to each organization) and the relationship of inputs to outputs. In determining the appropriate measures, one needs to consider the investment a firm makes into innovation in the form of resources and the environment it provides. One also needs to consider results achieved from that process and the impact these results have on the market value of the firm.

Innovation as an intuitive and creative process is a difficult process to measure. Innovation, which is considered an art, historically is measured in terms of financials or counts. Innovation, being a complex and unknown process, proves to be a challenge when defining clear and correlating measurements. The financial- and count-type measurements include product- or service-specific sales or revenue growth, and count-type measurements include items like the number of patents, trademarks, articles, and product or service versions produced. However, experience shows these measurements do not correlate to the innovation activity; therefore they should not be used as a business measure of performance.

In order to establish measures of innovation, understanding the innovation process first is a must. Corporations implement innovation through the network-centric, pipeline-fed, and opportunity-driven approaches. The network-centric approach, which is taught in colleges, is based on collaborative brainstorming. The concept is that more minds are better than one at a given time (without understanding the "why"). The pipeline model is driven by inventors who work in a research and development environment on a specific topic, explore new ideas and develop new products and services. The pipeline model, which is driven by chance or innate genius, is a somewhat common perception of the innovation process.

The opportunity-driven model is more representative of street-smart individuals who take an idea at the right time and the right place, devise a solution, know how to market it, and capitalize on their breakthrough. They also appear to be lucky, which is defined as an intersection of continual preparation and opportunity. Their success represents a once-in-a-lifetime windfall out of the blue sky (i.e., fortuitous occurrences). Another innovation process, which is a combination of collaboration and opportunity, is called the "open innovation" process and leads to products such as Linux and the Internet.

DIFFICULTY WITH CURRENT INNOVATION MEASURES

Such variants of the innovation process and its outcome are difficult to measure. Peter Drucker's process, detailed in his book, *Innovation and Entrepreneurship*, identifies various phases of innovation, including the phases of opportunity identification, analysis, acceptability, focusing on core idea, and leadership. The act of innovation, though, is still not clearly explained. Measuring innovation effectively is contingent on understanding details of the innovation process, its inputs and outputs, and its controls.

Measuring innovation is an important issue, as business growth and profitability in the knowledge age depend on innovation. Continual acceleration in innovation will sustain revenue growth, which will then fuel more innovation. Therefore, sustainable growth requires sustainable innovation, which requires that innovation be institutionalized and its output made predictable.

Today, corporations are skeptical of adding new measurements to the existing portfolio of measurements. The current measurements do not, however, get fully utilized through analysis for extracting business intelligence or continually creating new opportunities. Instead, too many companies end up measuring too many things for too little value. Currently a variety of dashboards and scorecards focus more on display instead of extracting intelligence out of the data. Such tools are more appropriate for data mining. The waste continues to pile up without extracting a proper understanding of the related processes as well as a planned application of the lessons learned from the measurements.

A similar approach is taken regarding the measures of innovation. Several institutions, corporations and consultants are developing measurements of innovation. Some are interested in developing innovation scorecards (e.g., www.petercohan. com, European Business School and Little, 2001) and innovation indexes (innovation radar or innovation dashboards). However, most of the measures lack a consistent definition of innovation and its elements.

As a result, innovation surveys are the most commonly deployed tools to determine an organization's readiness for innovation, its innovation capability, and innovation performance. Several players are trying to bite the innovation apple from different sides. Eventually all these measures will converge when we have a better understanding of the innovation process. Because the understanding of innovation is currently fragmented, however, so are the measurements.

Recently-conducted research shows that the correlation between various measures of innovation and its impact on the output is somewhat limited. Measuring innovation is therefore challenging, because current measures do not provide statistical analysis or relate the impact of innovation with any degree of confidence. Measures of innovation are not available for strategic planning because of the uncertainty associated with measures of the financial impact of innovation. Ultimately a lack of financial, organizational, and cultural structure around innovation exists. Research, however, has been done on the effects of innovation on workplace organization (Zoghi, Mohr & Meyer, 2005), corporate performance (Hensen & Webster, 2004; Rogers, 1998), organizational success (Editorial, Rgmag. com, May 2005), and organizational climate (www.cpsb.com, 2002).

CURRENT MEASURES OF INNOVATION

The climate of creativity and its effect on innovation was analyzed by The Creative Problem Solving Group to evaluate the link between climate and organizational innovation. The study included the following nine criteria:

- Challenge/Motivation
- Freedom
- Trust
- Idea Time
- Play/Humor
- Conflicts
- Idea Support
- Debates
- Risk Taking

The normal gap between an average company and an innovative company is about 25%, and an additional 40% gap exists for the above categories between the worst company and an innovative company. Analysis of the data shows that innovative companies outperform in the areas of risk taking, play/humor, challenge or motivation, and idea support. The most significant factor that differentiated an organization for innovation is risk taking. Innovative companies encourage risk taking by their employees. They create a culture of risk and reward in order to intellectually engage employees.

The PA Consulting Group (2006) identified nine dimensions, which the consulting group uses, to measure an organization's innovativeness:

- Committed leadership
- Clear strategy
- Market insights
- Creative people
- Innovative culture
- Competitive technologies
- Effective processes
- Supportive infrastructure
- Managed projects

Recognizing measures of innovation is a challenge because of the mismatch between the financial cycles and the long innovation cycle of concept to commercialization. Determining a real measurable output, in the timeframe when the cost is incurred, is difficult. Moreover, the output of the innovation process is sometimes dependent on many direct and indirect contributors in the organization. Arriving at predictable measures of innovation appears to be almost impossible. Some examples of innovation measures cited on the PA Consulting Group website include speed of development processes, competency metrics, and number of patents.

Tim Studt, in an editorial in *R&D Magazine,* explored the 'measures of innovation' topic. He suggests that measuring innovation is a do-able task with complexities of its own. The role of innovation has changed in the last 30–50 years, from the research and development (R&D) efforts to broader efforts in an organization. The measures of innovation, therefore, are shifting from primarily R&D spending to measures on tangible processes, product enhancements, and intangible investments. Because of the variation from product-to-product and organization-to-organization, as well as the subjectivity of an organization's innovation capabilities, any comparison of organizations using a set of measurements will be subjective. According to this editorial, key components of successful innovation include:

• Funding for innovation
• Trained and educated staff
• Collaborative environment
• Key individuals
• Corporate infrastructure
• Strategic planning

These factors consist of subjective and objective measures or components of innovation.

Innovation scorecards are another way to measure innovation. Innovation scorecards are a graphical display of measures and the status of an organization. However, the challenges with this method are the measures to include in the scorecard, and how to collect data for those measures, both internally within an organization and externally for benchmarking purposes.

In 1986, a group of industrial, academic and labor leaders organized to form The Council of Competitiveness (www.compete.org) to address the trends in U.S. competitiveness and to act as a catalyst to launch national initiatives for improving U.S. competitiveness. The Council developed a national innovation index for assessing U.S. innovation capacity. The index consists of the following four types of measurements (NII Report, 2004):

• The quality of the common innovation structure
• The cluster-specific innovation environment
• The quality of linkages
• Other measures

The variables that constitute the innovation index are patents, R&D personnel and expenditure, trade regulations, protection of intellectual property, investment in education, and university and private industry research participation. The index is utilized to assess the relative competitiveness of various developed and developing economies and to project emerging centers for innovation.

The Council launched the national initiative, similar to several European national initiatives, for improving competitiveness and raising the need for accelerating innovation. The Council's national agenda highlights three key components— talent, investment and infrastructure. Accordingly, elements of these key components include improving education, developing innovators, opening new frontiers, promoting entrepreneurship, rewarding risk-taking, protecting intellectual property, strengthening manufacturing capacity, and developing growth strategies.

The European Business School and Arthur D. Little jointly developed The Innovation Scorecard for creating value through innovation management. The scorecard measuring innovation performance is driven by the following elements:

- Innovation Strategy
- Organization Favoring Innovation
- Innovation Process
- Innovation Culture
- Resource Deployment

The European Union Regional Innovation Policy also developed a scorecard for ranking the strengths and weaknesses of a region against a number of criteria. The main elements of this innovation scorecard include:

- Employees in medium and high-tech employment
- Internet users per 100 inhabitants
- Business enterprise R&D
- Research infrastructure
- University research income
- University research strength
- Knowledge workforce
- Qualification level

Kellogg School of Management (Walcott, 2003) presented an Innovation Radar that is designed to create a holistic framework to visualize, diagnose, benchmark and improve the innovation process. An initial set of measurements incorporated in the innovation radar include:

- What (Offerings), Brand, Networking
- Where (Presence), Supply Chain, Organization
- How (Process), Value Capture, Customer Experience
- Who (Customers), Solution, and Platform

Benefits of such a representation of the corporation's innovation performance are to identify strengths and weaknesses in the innovation process in order to develop a sound innovation strategy. However, similar to the other measures of innovation, the following questions remain to be answered:

- What data to collect?
- How to collect data?
- How to analyze the data?
- How to interpret data?
- How to drive improvement?

Extensive research conducted at the University of Melbourne addresses issues with measuring innovation. Accordingly, the accuracy of innovation measurements is critical to assess the extent of innovation and its impact on economic and social well-being. A fundamental challenge in economic analysis is that static indicators are the basis for measuring dynamic processes. Paul H. Jensen and Elizabeth Webster of the Melbourne Institute of Applied Economic and Social Research (MIAESR) identified four specific dimensions to the problem of measuring innovation. They are as follows:

1. The innovation process may take years from concept to commercialization;
2. In the narrow sense of innovation, the novelty of products or services is difficult to benchmark, and the process measurements are difficult to adjust;

3. Time carries an important economic value for the innovation process. Therefore, innovation measures must have some way to adjust in value over time;

4. Much of the innovation activity is categorized as unobservable and is not reported in conventional methods.

The authors, Jensen and Webster, identified three main characteristics of innovation measures: Type of Innovation, Stage of Pathway, and Firm Characteristics. The measures of innovation in their reported research included patent applications, trademark application, design application, expert assessment, journal counts, and survey of managers. A challenge exists in identifying a complete list of innovation measurements.

An analysis of various innovation measures identified the main industries active in innovation, including manufacturing, wholesale trade, finance and insurance. This analysis involved examining several hundred firms and found no significant correlation between measures of innovation and the firm's size. However, data were collected using surveys, which did not appear to be a good measure of innovation.

PROCESS BASED MEASURES OF INNOVATION

Mark Rogers of Melbourne Institute at the University of Melbourne has attempted to establish measurements of innovation at the corporate level. Rogers also identified input and output measures of innovation, along with their descriptions and (more importantly) the source of data collection. The recent effort in collecting data for measures of innovation has led to difficulties in determining what data to collect. Should the data collected be data that are already available in public records (annual reports), data that can be requested (through surveys or interviews), or data that are yet to be collected? Some of the input and output measures are listed in Table 10.1, Sample of Input and Output Measures of Innovation:

Each measure of innovation has some validity, but none can be used as a stand-alone measure of innovation. However, combining various measures to develop an index of indicators

TABLE 10.1. Sample of Input and Output Measures
of Innovation

INPUT MEASURES	OUTPUT MEASURES
R&D	Introduction of new or improved products or processes
Acquisition of technology	Percentage of sales from innovative products or processes
Expenditures associated with innovative products or processes	Intellectual property
Marketing and Training expenditures	Firm's financial performance

Note: Rogers (1998)

must consider tangibles and intangibles, economic and non-economic measures of innovation-related resources, processes, deliverables, and value.

Based upon the current research, the innovation process has been a fuzzy one at best. In order to establish a set of working measures of innovation, one must identify common characteristics of the innovation process, their inter-relationships, and well-defined deliverables. Figure 10.1 shows the process of innovation using the 4-P model. The model illustrates that for an innovation process to be standardized, identifying inputs, in-process activities, and outputs must occur.

The inputs include elements such as information, tools used in the innovation process, the approach to innovation, and targeted innovation output. The target may be a service, a product, or a certain change in product or service characteristics. According to Gupta's Einsteinian Theory of Innovation (GETI) delineated in Chapter 7, the innovation depends upon resources, knowledge, play, and imagination. The resources may be headcounts, equipment, or the acquired knowledge itself. The process includes execution, incentives, recognition, collaboration, and research.

The SIPOC (Supplier, Input, Process, Output, and Customer) model can be used for analyzing the innovation process.

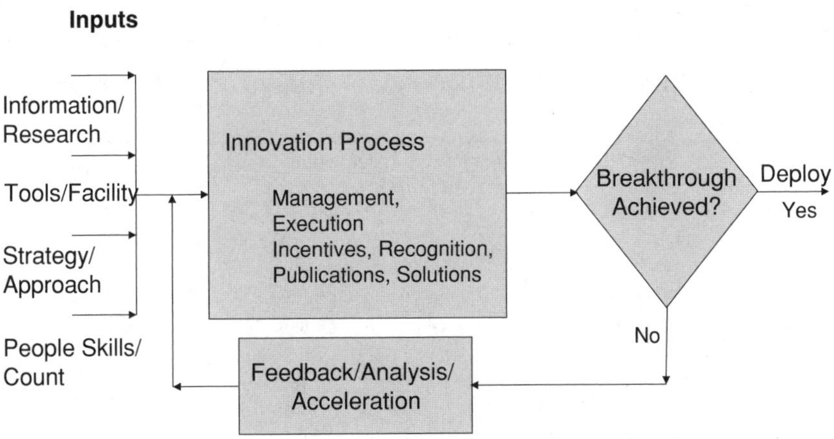

Inputs

FIGURE 10.1 Understanding Measures of Innovation

Table 10.2, SIPOC Analysis of the Innovation Process, shows various elements of the innovation process. Depending upon an organization's needs, these elements can become the measures of innovation.

The analysis of the innovation process shows many process steps and dozens of measures that can be used for monitoring innovation. The challenge is that people want to devise some magical measures of innovation that can tell the whole story and serve as predictors of innovation. Most management people would like to identify some measures, set targets, provide incentives, and start monitoring them. Even with a better understanding of the innovation process, a lot more thinking still needs to occur before selecting appropriate measures of innovation for an organization.

Given the current understanding of the innovation process, establishing an adequate and accurate measurement system right away is unlikely; instead starting an initial set of measures is a more appropriate way to begin measuring innovation. Figure 10.2 shows how a set of hierarchical measures are established to measure the effectiveness of the innovation process. Figure 10.2 also shows how people can get confused in devising a measure of innovation without fully understanding its components.

TABLE 10.2. SIPOC Analysis of the Innovation Process

Supplier (Source)	Input	Process	Output	Customers
Customer, Fundamental research, supply-chain	Demand from customers, Demand defined based on market research, Partner expectations	Establish targets for innovation	Solution for variety of selected and valuable demands	End users, marketplaces, businesses
Customer requirements driven strategic plan, Collaboration Tools	Identified domain I expertise or competence, Field of solution, Collaboration process	Teamwork with necessary knowledge base	Collaborative work, Superior output than that of individuals	Organization
Internet based access to research databases, Publications	Variety of information, Benchmarking information	Research the topic	Expanded understanding of the domain and related domains, topic of interest, Exploration capability, Applicable alternate sources of solution	Management, Innovators, Organization, Knowledge repository, Publications or Patents
Knowledge Repository, Innovators, Management	Alternate solutions, Internal capability	Make, "acquire" or "innovate" decision	Commitment to innovate, or acquire	Management, Innovators
Team members, Management, Suppliers, Collaborators	Knowledge, Resources, Environment, Methodology, Tools	Play to innovate	Good ideas, Crazy ideas, Funny ideas, Innovative ideas	Innovators, Team members
Management, Innovators	Culture for Creativity, Good ideas, Crazy ideas, Funny ideas, Innovative ideas	Develop alternate solutions classification,	Evaluation, Performance Trademark Office Patents, Publications	Organization, Patents and
Management, Organizational expectations	Alternative solutions, Organizational Expectations	Select a solution	Product, Service, Process, Platform	Organization, Marketing and Sales
Management, Organization	Performance measures, utilizing the solution	Verify the solution for economic value	Alternate applications, Customer review	Marketing, Sales and Customer
Management	Market leadership intent, Market demand, Resources ($), Target market, Need identification	Develop marketing plans	Market plan to achieve necessary market recognition, Customer's interest	Marketplace, Customers, End users

TABLE 10.2. (*continued*)

Supplier (Source)	Input	Process	Output	Customers
Management, Resources	Strategic sales plan, Supply chain relationships	Commercialize	Growing product sales, New customer relationships, Sales and Distribution channels	Organization, Society
Management, Process Owners	Implementation of Six Sigma Business Score-card or equivalent measurement system, Effective data collection	Monitor impact on business performance	Improving measures of innovation in Business Performance Index, Higher Customer Satisfaction	Shareholders, Stakeholders Assess impact on market capitalization
Inside or outside data sources	Market research, data collection	Assess impact on market capitalization	Improved shareholders' equity, Innovative corporate image	Shareholders, End users
Process owners, Industry sources, Management	Model of Innovation Index, Data collection	Measure Innovation	Innovation Index	Management, Marketplace

For example, if one is developing a process measure to ensure its effectiveness, one needs to look into its inputs, activities and outputs. If one is interested in developing an innovation index for an organization, one must consider factors such as variation between entities and key selected processes of measures. When developing a process measure, considering its effectiveness in producing the desired result and its relationship with the inputs and activities is essential. In other words, when establishing measures of innovation, establishing a clear objective and purpose for doing so is a must. Once the purpose is defined, and the scope of measures is established, then critical inputs, activities and outputs are identified. Based on the feasibility of data collection already existing or yet to be created, the aggregation of measures and their interpretation, communication, and resultant process adjustments must be thought through in order to select meaningful measures of innovation.

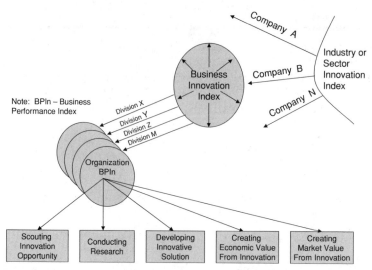

FIGURE 10.2 Sample Processes for Measures of Innovation

The Business Performance Index (Gupta, 2003) consists of seven elements and ten measures for monitoring business performance. The seven elements and ten measures are shown in Table 10.3, BPIn Measures, and include measures for idea management, sales growth, and employee recognition for exceptional improvement. All three of these measures are measures of an organization's innovation performance at various stages. The most critical resource with minimal financial impact is the effective intellectual participation of employees in developing innovative solutions as a way of doing work. This requires a robust employee idea management system which can enlist employee ideas daily, filter criteria continually, and escalate the value-driven ideas for improvement or innovation for further implementation.

CEO Recognition of Employees and incentives act as a catalyst for innovation, Idea Management is the process, and Sales of New Products, Services or Solutions is an output of innovation processes at a corporation. Combining these three (or similar) measures to develop a corporate index is simple and allows for the assessment of key aspects. The objective of innovation measures is to provide trends in performance and identify areas for adjustment to accelerate innovation.

TABLE 10.3. BPIn Measures

Six Sigma Business Scorecard Elements	BPIn Measures
Leadership and Profitability	**CEO Recognition of Employees for Exceptional Value Creation** Corporate Profitability
Management and Improvement	Managing Rate of Improvement
Employees and Innovation	**Employee Ideas for Improvement and Innovation**
Purchasing and Supplier Performance	Suppliers Performance (Sigma Level) Cost of Purchase Goods/Supplies
Operational Excellence	Aggregate Process Performance (Sigma Level) Cycle Time Variance
Sales and Distribution	**Sales for New Products, Services or Solutions**
Service and Growth	Customer Satisfaction

EFFECTIVE MEASURES OF INNOVATION

The question, then, is how to identify good measures of innovation. Victor Basili established a Goal-Question-Metrics (GQM) paradigm to identify process and product measurements in the software engineering environment. Accordingly, GQM utilizes a top-down approach to define the goals behind measuring the software process, and then utilize these goals to determine precisely what to measure. Using the GQM approach, measurements are a means to realizing the end but are not the ultimate end. Instead, measurements must be focused on specific goals, applied to all life cycle products, processes and resources, and interpreted based on the organizational context, environment, and goals. GQM helps in identifying organizational, need-driven, dynamic measurements in achieving business objectives. The GQM approach consists of the following conceptual, operational and quantitative level understandings of processes:

- Goal is defined as an intent or conceptual understanding in terms of products or outputs, processes or activities, or resources or inputs.

- Questions provide an operational understanding of measurements that can be used to assess realization of goals and objectives.

- Metrics represent the data that provide a quantitative understanding of the answers to the questions in assessing performance against goals. The data can be objective or subjective, or the object itself along with the viewpoint from which the data are taken.

Why does one measure innovation? What are the objectives behind measuring innovation? The objectives may be to establish a relationship between innovation and the market capitalization, predictability of the innovation process (or the rate of innovative products), to determine performance of the innovation process, to assess availability of necessary resources for the innovation process, or an overall assessment of the innovation process for an organization. Considering each question one by one requires a different approach for answering each question appropriately. Data needed to quantitatively answer each question is different, and thus the measurements for innovation are different.

Accordingly, in order to identify innovation measures, understanding the purpose of innovation, its environment, and the input, in-process, and output parameters is essential. Furthermore, the relationships between input and output innovation variables must be implicit. To determine measures of innovation, understanding the role of each process in creating the desired innovation is essential.

Figure 10.3 shows steps in the innovation cycle, starting with the demand for innovation to publicizing innovation. The demand for innovation may be internal as well as external; therefore, measuring the location of innovation demand is important, because demand will drive the innovation activities within an organization. If developing measures for the entire innovation cycle is desirable, then an aggregate measure or a set of measures for innovation is critical.

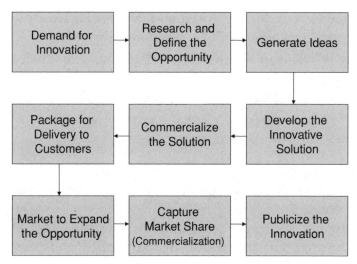

FIGURE 10.3 General Cycle of Innovation

With this understanding of measurements, an organization attempting to develop measures of innovation must clearly state its objectives before establishing the measures of innovation. Given the presence of a glut of measurements with no use in most organizations, an addition of nice-to-know measures is often perceived as "additional" work and not received well within the organization. Therefore, following is a list of steps used to establish measures for a process or an activity:

1. Define the **purpose** of innovation in the organization.
2. Establish expected **deliverables** (basic and specific) and their contribution to business performance, including growth and profitability.
3. Determine the **measures** of success of key deliverables.
4. Identify challenging **opportunities** for improvement in the innovation process.
5. List **activities** that must be performed to accelerate innovation.
6. Identify **input, in-process, and output variables** that are critical to the success of innovation in the organization. If

these variables are not monitored and managed effectively, the innovation outcomes will be adversely affected.

7. Determine the **data collection** capability of selected measures of innovation.

8. Establish **reporting** and communication methods, and monitor (levels and trends) critical and practical measures of innovation to drive business growth and profitability.

The above steps can facilitate the development of meaningful measures of innovation. Appropriately, the author has attempted to define innovation first before identifying its measures. Innovation is the application and commercialization of new ideas to products, processes or any other activities. Innovation means significant improvement can be realized through creation, collaboration or diffusion. Innovation can mean different things to different entities. Therefore, the measures will also vary from one organization to another organization.

FACTORS TO CONSIDER

Having understood the innovation process, measures of innovation, and elements of innovation indices, all that remains is actually implementing an innovation measurement system at one's organization. Simply copying another organization's measures of innovation is not sufficient, as they may reflect different priorities in performance and resource commitment. Table 10.4, Factors to Consider for Measures of Innovation, lists a variety of measures that can guide thinking in the right direction and facilitate development of appropriate measures of innovation. Good measures of innovation, being specific, measurable, and actionable, catapult the innovation process and produce significantly more innovative outcomes.

Current corporate measures do not include most of the measures of innovation. Being a new process evolving towards standardization, difficulties are expected in collecting data for various measures of innovation and benchmarking. However,

TABLE 10.4. Factors to Consider for Measures of Innovation

Industry Innovation Indicators	Business Innovation Index	Process Innovation Measures
Innovation Funding, including R&D	Resources—Funding, Culture of Risk Taking, Rewards, Tools	Excellence in Research, Innovation Management, Time allocation (%)
New Products, Services, or Solutions	Activities—Targets for Innovation, Process of Innovation, Extent of Institutionalization, Idea Management, Internal and External Publications, Knowledge Management, Internal and External Collaboration, Recognition	New Idea Deployment, Extent of Improvement or Change, Degree of Differentiation, Disruption or Innovativeness; Time to Innovate
Market Capitalization	Outputs—Patents, New Products, Services or Solutions; Sales Growth, Market Position or Ranking, Customer Perceptions	Rate of Innovation, Savings, Opportunities

institutionalizing and measuring the innovation process must start in order for innovation acceleration to occur in a measurable way.

TAKE AWAY

1. Innovation as an intuitive and creative process is a difficult process to measure.
2. The financials and counts type measurements do not correlate to the innovation activity; therefore they could not be used as a business measure of performance.
3. Corporations have implemented innovation through the network-centric, pipeline-fed, and opportunity-driven approaches.

4. Sustainable growth requires sustainable innovation, thus requiring that innovation be institutionalized and that its output be predictable.

5. Most of the measures lack a consistent definition of innovation and its elements. Since the understanding of innovation is fragmented, so are the measurements.

6. To identify innovation measures, one must understand the purpose of innovation, its environment, and the input, in-process, and output parameters.

7. Three good innovation measures include measures for employee recognition for exceptional improvement, idea management, and sales growth.

PART III

Institutionalizing Innovation

INNOVATION IN SERVICES

ALEXIS P. GONCALVES

Delivering services fundamentally differs from delivering a manufactured product because in services, the production, delivery and consumption by the customer frequently occur simultaneously. As a result, innovation in services can typically be categorized into Service, Process, Market and Business Model innovation. In addition, services more intensely focus on the customer. The proposed CI-3 framework classifies various approaches to innovation into Customer Intelligence, Customer Intimacy and Customer Innovation. This innovation framework can help organizations to identify, assess and select opportunities to add value to the firm and the customer. Being well-connected with its customer and having an in-depth understanding of customer needs is most important to a service firm.

How do service organizations innovate? How do they view their innovation activity? How do customer intelligence and customer intimacy drive innovation and new value creation in services? More specifically, how do these driving forces combine to increase innovation effectiveness? And how should an effective innovation system in a services organization be structured?

In addition to discussing those questions, this chapter presents a general framework for innovation effectiveness in services. The framework is called CI-3 and stands for Customer Intelligence, Customer Intimacy and Customer Innovation. The framework is based on practical experience in implementing methods to improve innovation effectiveness in service organizations, as well as a review of a large amount of existing empirical research on innovation. The CI-3 framework will help service organizations to break free from industry orthodoxies, align people around specific innovation opportunities, and use smart discovery and innovation methods to generate value for customers.

The CI-3 framework is applicable to service industries in general. Even if the service industries are different from one another, practical experience and empirical research indicates that some common characteristics of the innovation processes are present due to the specific nature of service production that is common to all service industries.

SERVICE ORGANIZATIONS DO INNOVATE

Services have been considered in some research as an appendix to manufacturing, a residual sector, or at least a sector lagging behind the manufacturing sector in the form of low productivity, low capital intensity, weak qualification levels, and low innovation activity. This statement is not true at all. Services are the name of the game in today's economy.

Services represent about 80% of the U.S. gross domestic product and between 60% and 80% of the GDPs of the rest of the world's advanced economies (Chesbrough, 2005). Excelling at service management is a priority, not only in the USA but also abroad in places like Asia, Europe and South America (Goncalves, 1998). Companies like General Electric, Xerox, and IBM that are seeing their own businesses shift from products to services are acutely aware of this trend. At IBM, for example, more than half of total revenue now comes from services.

Service firms do innovate, but the innovations often take other forms than they do in manufacturing, and they may be organized differently. The different forms of service innovations are, to some degree, related to the specific form of service production. Specific characteristics of service production are not delineated here, as they are treated intensively by other authors (Levitt, 1972, 1976; Lovelock 1983, 1996; Gummesson, 1988; Sundbo, 1994); however, briefly repeating the facts with regard to service production that are most relevant to innovation are highlighted here.

In service industries the product is not always perfectly "formatted" and codified, (i.e., precisely determined a priori). Each service transaction may be considered unique as far as it is produced on demand (tailor-made) in interaction with the client or as a response to a specific, non-standardizable problem, and in different environments. Client participation (in various forms) in the production of the service may be the most basic characteristic of service activities, particularly knowledge-intensive ones.

Various concepts have been developed in order to account for this client involvement: co-production, servuction, service relationship, the moment of truth, and prosumption (Eiglier and Langeard, 1991; Carlzon, 1993; Tapscott, 1996). At the interface between the service provider and its client, different types of interaction are occurring. Different types of elements are being exchanged, including information and knowledge, emotions, and verbal and gesture signals of civility. This interaction also expresses power struggles, domination, and reciprocal influence relationships.

The distinction between product and process is widely accepted in the case of manufacturing goods. The same is not true of services where the product mostly cannot be separated from the process. Here, the term "product" frequently includes a process: a service package, a set of procedures and protocols, or an "act."

In services, simultaneous production and consumption of the service occur. For example in a restaurant, production and consumption of the service are taking place at the same time. The understanding of consumption as a "process" prompts many service organizations to apply *Lean Consumption Principles* and treat consumption not as an isolated moment of decision about purchasing a specific product, but as a continuing process linking many goods and services to solve consumer problems (Womack, 2005; Swank, 2003).

THE CONCEPT OF INNOVATION IN SERVICES

The notion of innovation in services is difficult to define and, in particular, difficult to delimit. The question of whether or not something is an innovation is sometimes difficult to answer. Several definitions of innovation have been forwarded in the literature. "Innovation in services is (a) change in things (products/services) which service organizations offer and (b) change in the ways in which they are created and delivered" (Goncalves, 2004). This definition encompasses the notion that in services, innovation is also a change in transaction or operational processes.

One could, as Schumpeter (1934) did, add that innovation also means changes in organizational forms, in the use of input or in the firm's marketing approaches. In the case of services, the latter could include innovations in the delivery system, which is extremely important in services. Furthermore, innovation is generally seen in the literature as technological change, but it could also be a non-technological change (i.e., a social or organizational change). The latter interpretation is relevant as it relates to innovation in services.

A further issue is how big the change should be in order to be considered an "innovation." Should it just be any small renewal, or should it have a certain degree of change? Obviously, a different challenge is involved in creating a new insurance policy versus something like the identification and development of a viable new business concept. All degrees of innovation count, and the matter to consider is how to define the right degree of innovation for any one firm.

Two elements of analysis need to be considered by any services organization. One is recognizing a firm's competitive condition so that it can pursue the right degree of innovation. Once a firm makes that choice, the second element to consider is using methods adequate for the task. More challenging innovations, for example, demand better teams, better approaches to customer insight, and more resources.

Another key concept associated with innovation is reproducibility; how reproducible must the innovation be? In services, a change may be a new way of solving the problem for a client, which will only be done once and never be repeated. Is such an example representative of a true innovation? One can suggest a minimalist definition of innovation that includes all such situations, but such a definition will very easily lead to the situation where innovation is about everything, since we live in a rapidly-changing world where everything changes.

TYPES OF INNOVATION IN SERVICES

Innovation in services can be a new service product, a new procedure for producing or delivering the service, or the introduction of a new technology. Since service in most cases cannot be stored, it must be produced at the moment of consumption (Grönroos, 1990; Eiglier and Langeard, 1991). This means that the procedure cannot be completely separated from the product, which leads to the conclusion that changing the product without changing the procedure is difficult.

Thus, service innovations are generally broad in the sense that they imply a change of many elements in the production process and the product simultaneously. From the service management and marketing literature (Normann, 1991; Gummesson, 1994), one can conclude that the customers, and the solution of their problems, are so important that innovations in service firms must be explained from the customer side (i.e., "pull oriented"). Innovations in services are customer oriented, but they are also often developed from ideas within the service firm (i.e., "push oriented").

The ideas may have evolved from the interaction between service personnel and customers, but the ideas are not directly presented by the customers nor do they directly answer one single customer's concrete problem. The degree to which the innovations are customer-determined is different in different service industries or segments. Within standardized services, such as cleaning or retail banking, the innovations are less customer-determined than within advisory services, which are less standardized and much more individualized to the single customer. In advisory services, for example, innovation is often an interaction process between the service provider and the customer, where the development of the innovation is taking place within the customer firm. Both parties may learn from the innovation process and exploit it. In other words, the customer firm benefits by having solved some problem and the service provider can generalize the solution and sell it to other customers.

Another distinction between manufacturing and services is that innovation in services is more often a nontechnological or social innovation (as opposed to a technological innovation). Innovation in services should not be understood from a narrow technology-determined viewpoint. Innovation in services may be the creation of new knowledge or information, or new ways of handling things or persons, which are just new types of behavior by the service personnel. Innovation in services may also be small adjustments of procedures (and thus incremental and rarely radical). The development time for innovation in services may be relatively short if no need for research or collection of scientific knowledge is present (Shostack, 1984).

Based on the characteristics of service previously discussed, innovation in services can be categorized into four types:

- *Service innovations:* The creation of new service products.
- *Process innovations:* The renewal of the prescriptive procedures for producing and delivering the service. The process innovation can be divided into two sub-categories: innovations in operational processes ("back office") or in delivery processes ("front office").
- *Market innovations:* The creation of innovations in marketing and commercialization (e.g., finding a new market segment, entering another industry and its market). As an example, consider the case of a large retailing chain that starts selling financial services (Zellner, 2005).
- *Business model innovations:* The creation of a new business concept in which the combination of the previous three types of innovation (i.e., a "new service," a "new process" and a "new market" forming a new business model) are present. Examples of innovation in the business model include Dell, FedEx and eBay.

In knowledge intensive business services, another type of service innovation—the "one-timer" innovation—can be considered. The "one-timer" innovation is the interactive construction of a solution to a particular problem posed by a client. This type of innovation is co-produced by the client and the service provider. It is not reproducible as such, but it may be reproduced indirectly through its codification, the formalization of part of the experience, and the development of innovation recipes.

The typology of innovation previously described can be adjusted. For example, market innovations can be divided into four subtypes as follows: (a) "brand innovation"—how the offering's benefit is communicated to customers, (b) "customer experience innovation"—how an overall experience is created for customers, (c) "channel innovation"—how offerings are connected by the firm to its customers, and (d) "segment innovation"—how the firm enters another industry and its market.

Some examples of innovation in services are:

- New financial products or insurance policies
- Development of service management systems
- e-Commerce networks with high-volume customers
- New advisory services in health care and well-being
- Distribution of special goods: i.e., transport of ready-to-eat "sushi"
- Premier pet program for frequent flyers traveling with their pets
- Introduction of automatic alerts to customers by e-mail, PDA or phone

THE CI-3 FRAMEWORK

Today managers in the services industry are under tremendous pressure to innovate and create new value for their customers as profit margins shrink and competition intensifies. Yet the most common responses to these pressures (Six Sigma, CRM technology, price cuts, outsourcing, cost reduction, process reengineering, and so on), while critically important, cannot solve the problems of margin pressure alone.

At the same time, major discontinuities in the competitive landscape brought about by rapidly converging advanced technologies, a changing and empowered consumer, and an increasingly networked society are calling into question the basic conception of customer value and the processes that lead to its creation. Managers are discovering that neither value nor innovation can be successfully generated through a product-centric or service-centric orientation alone.

New perspectives are urgently required to meet both the current and potential need for sustainable innovation and profitable customer value creation. As service organizations focus on identifying powerful new sources of customer value and innovation, they need new strategies that rapidly achieve business and customer objectives for mutual profitable returns.

The CI-3 framework (see Figure 11.1) helps service organizations to identify, assess and select innovation opportunities (i.e., value-creating opportunities) and to quickly set a course for change from a diverse range of approaches. The CI-3 framework is composed of three CIs and nine approaches as follows:

1. Customer Intelligence (CI)—A company's approach to build intellectual capital about its customers and use those assets to make strategic decisions about innovation, customer relationship management and service offerings. This element of the framework is comprised of three approaches:
   (1) Brand Asset Monitoring
   (2) Customer Satisfaction and Loyalty Measurement
   (3) Touch-Points Data Management
2. Customer Intimacy (CI)—A company's approach to gain intimacy with its customers and establish a relationship that enables co-creation and co-design of new services and products. This element of the framework is comprised of three approaches:
   (4) Customer Blending
   (5) Customer Mind Mapping
   (6) Customer Ecosystem Mapping

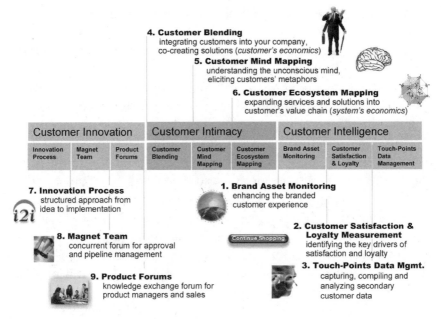

FIGURE 11.1 Nine Approaches Deconstructed

3. Customer Innovation (CI)—An organization's governance system to leverage all the intelligence acquired and intimacy developed around customers to innovate and deliver new, superior customer value. This element of the framework is comprised of three approaches:

    (7) Innovation Process
    (8) Magnet Team
    (9) Product Forums

Each one of the three CIs and the nine approaches are described in the following sections. Whenever appropriate, references are made to a company that is advanced in its approach, capabilities and competency with respect to the specific approach of each framework element being described.

CUSTOMER INTELLIGENCE

A company's approach to build intellectual capital about its customers and use those assets to make strategic decisions about innovation, customer relationship management and service offerings.

Developing intellectual capital about customers and the marketplace is of critical importance when identifying new value, balancing resources and building closer relationships with customers. Customer intelligence entails learning about customer needs, the influence of technology, competition, and other environmental forces, and then acting on that knowledge in order to become competitive.

Customer intelligence is the key that unlocks the value chain of a services organization. It allows organizations to identify critical-to-customer requirements and then take action on this intelligence. Customer intelligence creates a unified conversation that spans marketing, sales, and service and reaches across media and channels. It has three basic components; each component plays an integral role in (a) creating intellectual capital about customers and the marketplace, and (b) sustaining the development of customer intelligence. The three basic approaches to develop customer intelligence are as follows:

(1) Brand Asset Monitoring (develop intelligence about how to enhance the branded customer experience).
(2) Customer Satisfaction and Loyalty Measurement (develop intelligence about what the key drivers are of customer satisfaction and loyalty).
(3) Touch-Points Data Management (develop intelligence about customers by capturing, compiling and analyzing secondary customer data, or learning from every customer interaction).

Brand Asset Monitoring

Brand assets (or liabilities) are linked to a brand's name and may add value to (or subtract value from) a product or service. These assets are grouped into four dimensions: brand awareness, perceived quality, brand associations, and brand loyalty (Davis, 2002). These four dimensions include all that people feel and think about the brand as a result of direct experience, word-of-mouth, moments-of-truth with the brand, and the brand's marketing activities. For that reason, brand asset monitoring is vital to the success of the brand. It enables brand owners to see where the brand's strengths and weaknesses lie and what forces drive these strengths and weaknesses, which in turn points to the nature and level of innovation needed to fulfill the brand's potential.

By understanding the strength of the customer relationship with the brand, customer intelligence can start to build. One can also gauge how vulnerable the brand is to new entrants or to short-term promotions, as well as what level of innovation is needed for the customer experience to increase brand loyalty.

Several proprietary models based on worldwide market research have been developed to monitor brand asset and evaluate opportunities for brand innovation. These models are grouped into two categories: "consumer oriented" model or "business performance measures" model (Davis, 2002). The most-used "consumer oriented" model is based on consumer perceptions using a consumer questionnaire that measures four main areas:

1. Differentiation: How distinctive is the brand in the marketplace?
2. Relevance: How relevant is the brand to the consumer?
3. Esteem: How highly does the consumer regard the brand?
4. Knowledge: How well does the consumer understand what the brand stands for?

Scores against the "differentiation" and "relevance" dimensions are multiplied together to produce a measure of *brand strength*. Scores against the "esteem" and "knowledge" dimensions are multiplied together to produce a measure of *brand stature*. This approach concentrates on the consumer at the expense of more business-oriented measures, such as market share or sales trends.

No one right way exists to evaluate brand assets and identify the level of innovation required. Rather, the various dimensions need to be gathered, reviewed, and prioritized in a structured brand-monitoring process and considered as a whole in the evaluation of the brand assets. The outcome of the brand asset monitoring process should be the identification of opportunities to innovate and improve the total customer experience.

Today, service firms are only slowly realizing that the customer experience is behind brand value. The routine interactions and/or transactions that occur between the customer and the service firm (a.k.a., moments-of-truth) highly impact brand image and loyalty. Customers always have an experience when they interact with an organization. They consciously and unconsciously filter a barrage of "brand images" and organize them into a set of impressions, both rational and emotional. Anything perceived or sensed (or recognized by its absence) is a branded experience. If you can see, smell, taste, or hear it, it is a branded experience. Services produce branded experiences, as does the physical environment in which they are offered.

The employees are another source of branded experience. Each branded experience carries a message; the composite of branded experiences creates the total customer experience. Many large service firms spend nearly 50% less on activities related to the customer experience than on advertising. This imbalance stems from years of misunderstanding the role of the brand, believing that it must be advertised, and viewing it as an expense to be managed quarterly.

In the financial services industry, some banks are breaking away. Banks with high-tech or coffee boutique atmosphere are using customer intelligence about their brand assets and investing heavily in customers' retail experience (Swan, 2003:Conley, 2005; Moore, 2005; Salter, 2002). Banks have tried too hard in the past to reposition their brands without delivering on the promise. For example, many banks promise convenience and then leave consumers frustrated and underwhelmed with their actual experience. Suppose a consumer must choose between two banks. The one with the more valuable brand will not automatically win. Instead, the consumer will react to an intricate set of emotional and pragmatic concerns: Is it convenient? Is it the same bank a friend uses? Do branch employees explain the checking account options clearly? Does the bank understand my needs?

Customer Satisfaction and Loyalty Measurement

The practice of regularly measuring customer satisfaction and loyalty is vital to the development of intelligence about customers and the identification of opportunities for innovation. Service organizations that have mastered this practice find (a) that a relationship between satisfaction and behavior exists, and (b) that the relationship is not linear. Rather, the "satisfaction-loyalty curve" appears to be nonlinear and asymmetric (Reichheld et al., 1996). In other words, the impact of changing customer satisfaction on repurchase intent, customer loyalty, price sensitivity, and so on is greatest at the extremes (i.e., extreme dissatisfaction or extreme satisfaction), with those attributes that contribute to dissatisfaction—"must-be-quality" attributes—and those attributes that contribute to high satisfaction—"attractive-quality" attributes (Walden et al., 1993).

In the middle part of the curve, however, changes in customer satisfaction are unlikely to make a difference in customer loyalty. In fact, many service firms discover that customer satisfaction does not noticeably impact behavior except when customers are exceptionally dissatisfied, and hence more motivated to look for alternatives and switch providers, or when customers are exceptionally satisfied (often at a point where the firm has achieved some type of "service excellence" threshold) to the point where their "delight" translates into loyalty.

Service firms must realize that satisfaction indexes reveal only one of several components that drive perceived value and customer behavior, explaining why incremental changes in satisfaction may not change customer behavior at all. For example, contexts occur in which relatively unsatisfied customers may stay with a provider despite their lack of satisfaction (if, for example, they do not think competing offers are better or provide greater value or it is too difficult to switch to them). Alternatively, a customer who is very satisfied with the service she receives at her local bank may nevertheless shop elsewhere for her jumbo mortgage to get the best rate. She may purchase her investment products from a third provider if she perceives that firm to be more sophisticated and better at providing advice. Furthermore, she may open an account with a fourth provider because it is convenient to her office, and she believes it will deliver the same level of service as her existing provider.

The most compelling value of measuring customer satisfaction and loyalty is to enable an organization to (a) conduct a comprehensive analysis of the customers and the marketplace, and (b) prioritize innovation opportunities that are linked to loyalty drivers. The following are examples of service organizations that are using customer satisfaction and loyalty measurements to build customer intelligence, become innovative, and deliver new value for their customers.

A bank, based in the Midwest, has developed a sophisticated system to track several factors involved in customer loyalty and satisfaction. Once driven strictly by financial measures, it now conducts quarterly measures of customer retention: the number of services used by each customer, or *depth of relationship*, and the level of customer satisfaction. The strategies derived from this information helped the bank identify opportunities for innovation and new value creation. This explains why the bank has achieved a return on assets more than double that of its competitors (Heskett et al., 1994).

Every month, a car rental company polled its customers using just two simple questions—one about the quality of their rental experience, and the other about the likelihood that they will rent from the company again. Because the process was so simple, it was fast. That speed allowed the company to publish ranked results for its 5,000 U.S. branches within days, giving the offices real-time feedback on how they were doing and helping the corporation to identify opportunities for innovation and new value creation (Reichheld, 2003).

A captive leasing company, developed multi-purpose customer listening tools as well as satisfaction measurement instruments. Surveys are customized by Division (according to customer type) and include User, Dealer, and Competitive Surveys. The company has twice raised the bar for reporting satisfaction results. Results are aggregated into the Customer Satisfaction Index (CSI), a multi-attribute index used to evaluate satisfaction levels. Factor analysis determines satisfaction with core processes. Multiple regression analysis determines the significance each factor contributes to overall customer satisfaction. Results are segmented and analyzed by market segment and process. Process Owners and Six Sigma teams use the data to identify opportunities for innovation and increase customers' perception of delivered value (NIST, 2003).

Touch-Points Data Management

Brand Asset Monitoring and *Customer Satisfaction and Loyalty Measurement* surveys are good approaches to listen to the voice of customer and identify opportunities for innovation. However, they are not enough. Service firms must respond, and in some cases make changes (i.e., innovate), when what they hear indicates problems or opportunities. Touch-points data management means building customer intelligence based on readily-available customer information that is scattered across different channels or touch-points.

Service firms too often do not make use of data they already have that could be used to identify opportunities for innovation. Touch-points data management is a useful metaphor for sifting and sorting customer information warehoused in several databases. The purpose of touch-points data management is to spot trends, relationships, and other nuggets of information and customer insights in order to identify better opportunities to innovate. Databases are the primary tool used in managing customer relationships, an activity or program sometimes referred to as *customer relationship management* (CRM). CRM programs started out as tools to help the sales force keep track of customers and prospects but, over time, CRM has become more strategic and customer focused.

Touch-points data management is the area where many service firms, multi-line firms in particular, tend to stumble. The reason for this is that their sales, marketing, operations and service activities are "siloed" around products rather than around customers who are critical to the firm's success (Peppers et al., 2004). As a result, customer intelligence is fragmented and limited, and the ability to understand the total customer experience is compromised.

In order to both build customer intelligence through touch-points and identify opportunities for innovation, service firms must conduct an extensive secondary data analysis and search for existing customer information across the entire organization (i.e., sales, marketing, customer service, call centers, after-sales support units and any other customer-touching group within the firm). In those areas/units of the firm, customer information may be stored in different formats, media like electronic databases (e.g., CRM systems like Siebel, SAP, Peoplesoft), or even folders and log-sheets. This type of analysis may use special software programs to periodically scan a customer database and indicate when an above-average number of complaints, inquiries or compliments are received within a certain period. The critical number will vary depending on the size of the customer base and the product or service category. Note that the customer database should capture all interactions (not just sales) receiving input from customer contacts handled by any customer touch-point in marketing, customer service, call centers, after-sales support units, and any other customer-touching group within the firm.

Excessive inquiries about a service or subject, such as "How do I calculate the interest rates paid on my mortgage contract?" or "Have my late fee charges been reversed already?" indicate either that more information needs to be made available to customers, or that the service needs to be redesigned so that it stimulates fewer questions about its use. An auto-insurance company, ran an exception analysis of its database and found that it was receiving thousands of calls a month from potential customers saying they would like to know how its rates compared with the competition. After discussion within its senior management team, the company decided to embrace "information transparency"—a policy of sharing with its customers (or prospects) information about prices, costs and service.

The company's "1-800-" service, for example, quotes its own rates to potential customers along with the rates of competitors, even if those rates are cheaper (Salter, 1998). Some people do not believe the company does this; they think it is a gimmick. This strategy, however, is part of its "information transparency" policy. The company may lose some customers who opt for lower rates, but management believes that transparency will retain remaining customers, build loyalty and establish a feeling of trust for the company. In an industry so notorious for high prices, bloated bureaucracies, and poor service, "information transparency" is an important brand asset.

When above-average numbers of compliments are recorded in a given period, the company should examine them to see whether they relate to some benefit of the service that could be further leveraged in the brand's marketing communication messages. Compliments should also be forwarded to the operations and service delivery areas as a way of recognizing good work and praising employees. Morale boosting improves customer service and lowers the number of service defects.

All services organizations, no matter how good they are, receive complaints. What is critical, however, is to note when too many are being received about a particular service or product. Negative exception reports act as an early warning system. Complaints indicate either that the company's services or methods of handling customers are faulty, or that customers' expectations are not being properly managed. Managing expectations is one of the primary responsibilities of marketing communications.

Touch-points data management can be used at each stage of a brand relationship: acquisition of customers, relationship deepening (up-sell), relationship broadening (cross-sell), and retention and reacquisition (recovery of lost customers). Probably one of the most critical stages for service firms is the retention and reacquisition stage. Customers leave a brand for any number of reasons. Retention and reacquisition programs can minimize the ultimate loss but require good listening and learning skills.

A credit card company has adopted an interesting approach to minimize attrition. Not only has it developed a strong analytical capability to predict customers' behavior and anticipate defection, but its senior executives learn from defecting customers. Each executive spends four hours a month in a special "listening room" monitoring routine customer service calls as well as calls from customers who are canceling their credit cards (Reichheld and Sasser, 1990). This approach shows actual leadership and visible sponsorship for touch-points data management.

Customer Intimacy

A company's approach to gain intimacy with its customers and establish a relationship that enables co-creation and co-design of new services and products.

As described before, customer intelligence is the starting point to build intellectual capital about customers and identify opportunities to innovate and improve the customer experience. However, although customer intelligence is a necessary element for innovation, it is not sufficient. A company needs to go beyond customer intelligence and achieve customer intimacy (i.e., close and trusting relationships to increase understanding about customers' needs). The word "intimacy" stems from the Latin *intimatus*, to make something known to someone else. In its original meaning, intimacy did not mean emotional closeness, but rather a willingness to pass on honest information.

Customer intimacy is fostered by a market orientation. Market orientation entails learning about customer needs, the influence of technology, competition, and other environmental forces and acting on that knowledge in order to become competitive. Accordingly, market orientation is an organizational learning capability consisting of cognitive associations (e.g., shared beliefs, values and norms) and behavioral outcomes reflecting these cognitions (Kok et al., 2002).

In this regard, in order to develop customer intimacy, service firms must become learning organizations (i.e., firms that have mastered organizational learning). Organizational learning is valuable to a service firm and its customers, because it helps support the understanding and satisfying of customers' latent needs through new products, services, and ways of doing business. Organizational learning makes it possible not only to create products ahead of competitors, but also to create them before the recognition of an explicit customer need.

According to Peter Senge, who popularized the concept (Senge et al., 1994), organizational learning involves two distinct types of behavior: adaptive learning and generative learning.

Adaptive learning focuses the organization on adjusting to serve the present market. With a continued focus on the present market, however, these core capabilities can dominate, and thus constrain, the direction and development of the firm. They can become core rigidities that inhibit innovation.

In contrast, generative learning requires an organization to challenge its own assumptions about its mission, customers, competitors, and strategy. If a company can look at its environment beyond its familiar assumptions, it may be able to discover new directions and new possibilities, and thus create new innovative services. Adaptive and generative learning are key elements for the development of customer intimacy. Following are the three basic approaches to foster learning and enhance customer intimacy:

(1) Customer Blending (integrating customers into the company and co-creating solutions to improve *customers' economics*)
(2) Customer Mind Mapping (understanding the unconscious mind of customers and eliciting customers' metaphors)
(3) Customer Ecosystem Mapping (expanding services and solutions into customers' value chains to improve the *system's economics*)

Customer Blending

The logic of customer blending rests on information asymmetry (or independency), because the "need" information (what the customer wants) resides with the customer, and the "solution" information (how to satisfy those needs) lies with the service provider. Implied in this is the assumption that the necessary information for reducing uncertainty concerning the new service development process exists and is possessed by the customer, and that service development is only a matter of finding where the required information is located and communicating it from where it is to where it should be.

Customer blending emphasizes the interaction with customers or *lead users* as an effective approach to innovation (i.e., a way to achieve a more favorable cost/time service development curve and to reduce uncertainty that usually surrounds the innovation process). Blending with customers or *lead users* is a crucial factor in the process of getting the innovation developed and accepted in the market. Customers or *lead users* are sources of information and knowledge, and interacting with them can enhance service concept effectiveness.

With regard to customer involvement in firms' innovation processes, researchers have identified five roles that these customers can play: customer as "resources," "co-producers," "buyers," "users" and "products" (Prahalad et al., 2002; Lundkvist et al., 2004). Whereas the first two roles are at the input side of the value creation process, the last three roles take place at the output end of it. Blending with customers or *lead users,* to innovate is a practice that may have different names. At one company it is called "dreaming sessions" (GE, 2004), at Citigroup it is called "customer blending sessions," and other common names are "customer lab sessions," "customer dialogue sessions," "customer symbiosis sessions," "opening the kimono to customer sessions," and "customer scenario mapping sessions" (Seybold et al., 2001; Sharma et al., 2002).

To be effective and create new value, the innovation process should be an interaction with different external actors, particularly with customers (*lead users*) but also with suppliers, sub-contractors, distributors and other members of the service value-chain. Traditionally, service firms have innovated by sending out market researchers to discover "unmet needs" among their customers. After the research is complete, the firm decides which ideas to develop and hands them over to a project development team. Studies suggest that about three-quarters of such projects fail (Von Hippel, 1999, 2002).

Harnessing customer innovation requires different methods. Some service companies, instead of taking the temperature of a representative sample of customers, identify the few special customers who seek innovation. Such customers are referred to as *lead users* and are perfect partners with which to blend and co-create solutions. At one company's healthcare division calls them "luminaries." They tend to be well-published doctors and research scientists from leading medical institutions, says the company, which brings up to 25 luminaries together at regular medical advisory board sessions to discuss the evolution of its technology. The company then shares some of its advanced technology with a subset of luminaries who form an "inner sanctum of good friends." The company's products then emerge from collaboration with these groups (Von Hippel, 2005).

An office-supplies retailer, found its luminaries by holding a competition among customers to devise new product ideas. The competition, named "Invention Quest," is a national search for the next great office product. In 2004 alone, the retailer received 8,300 submissions, out of which the winner was a gadget called "WordLock," a lock whose combination is formed by letters in easy-to-remember words rather than numbers (McGregor, 2004).

At a business unit at Citigroup, the *lead user* project team is formed by relationship managers and product heads. It uses two basic information-transfer methods. First, it interviews with and site visits to individual *lead users* that it has identified. Usually they are directors or managers within the finance department of the client organization (e.g., CFOs, Directors of Finance, Treasurers, Cash Management Heads, Trade Finance or Trade Services Heads).

Second, it invites a few *lead users* (six to eight), who appear to have very promising ideas and insights, to participate in a joint problem-solving workshop with members of their multidisciplinary *lead user* project team, which includes not only relationship managers and product heads but also managers from operations, risk management, and technology. Such workshops typically run for one to two day(s). During that time, lead users and the multidisciplinary *lead user* project team all join in the problem-solving work of designing one or more potential financial solutions that precisely fit the client organization's needs.

The trend toward blending with customers for innovation can be seen across the entire services industry, from software and information services to health care, consulting and banking. The more frequent and intense the communication among players (i.e., firms, customers, *lead users* and other members of the service value-chain), the more likely substantial innovations are to occur. Obviously, adopting appropriate measures that promote a trust-based culture between the firms' new service development teams and customers requires a cultural shift from a "we-know-best" attitude to seeing product ideation as the joint outcome of a customer blending process.

Customer Mind Mapping

Customer mind mapping is important to innovation, because the methods and approaches used to gain customer intelligence are mainly focused on satisfying expressed needs of the customer. This information is typically gathered by using verbal techniques, such as focus groups and customer surveys, to gain understanding of the use of current products and services. The problem is, however, that those techniques tend to result in minor improvements rather than innovative thinking and breakthrough services/products. This problem arises because customers have trouble imagining and giving feedback about something that they have not experienced. Organizations simply cannot access, understand, and meet latent customer needs (or "unarticulated needs") by using surveys and interviews. The incapacity of traditional market research techniques to anticipate latent customer needs has implications for innovation in services and new service development.

Latent needs (or "unarticulated needs") can be defined as what customers really value, or the products and services they need, but have never experienced or would never think to request. Latent needs (or "unarticulated needs") are needs that customers have not yet found a way to express—often because they are very novel or rapidly-evolving—but that they would be very pleased to have solutions to nonetheless.

An example of a latent need is this idea that occurred to a customer as she was walking towards an ATM to get her account balance. She passed by a fast food restaurant, and she heard someone's phone beep receiving a SMS. Then she realized that it would be very handy to get the balance on her account through the mobile phone once a day, and that the beep could be replaced by the restaurant's jingle. In that way she would be in better control of her money situation, and the fast food company could sponsor and pay for the service, as they would receive more advertising. This idea was an expression of the latent need of "getting information in real time."

Service innovations are found to be lagging behind in the area of innovations driven by insights into novel, latent, and unarticulated market needs. In order to uncover latent needs, recent findings stress customer participation in the development process or observations of customers in real action (Thomke, 2003). In this regard, understanding the mind of the customer and how the customer thinks is critical.

In 2003, Dr. Gerald Zaltman, a professor from Harvard, published a book explaining how the brains, minds, and memories of consumers work, and how service managers can effectively leverage that information in their innovation and new service development processes. He discusses the conscious and unconscious mind and how they work together to develop the metaphors and stories that drive consumer behavior (Zaltman, 2003). According to Zaltman, many managers handicap themselves with limiting views about consumers, such as:

- *Consumers think in a well-reasoned or rational way.* Consumers rarely assess benefits attribute-by-attribute and consciously balance the pros and cons of buying. The selection process is largely affected by emotion, the unconscious mind, and the social and physical context.

- *Consumers can readily explain thinking and behavior.* Ninety-five percent of thinking happens in the unconscious mind. Verbal explanations after-the-fact attempt to make sense of behavior, but they rarely explain what controlled it.

- *Consumers' memories accurately represent their experiences.* Memories are not always accurate representations of what happened, and they change over time.

- *Customers think in words.* The words expressed in surveys and focus groups only come after a person consciously chooses to represent unconscious thoughts out loud.

Based on the above misconceptions, one can sense that surveys, questionnaires and focus groups fail to get behind the curtains of consciousness (i.e., where motivations, perceptions and decisions originate). This disadvantage of these tools can prove fatal for an innovation program, because at least 95% percent of mental activity that leads to perceptions, thinking and decisions takes place outside the conscious mind. Traditional innovation practices and marketing research largely ignore the contents of the unconscious mind. Lacking an understanding of how minds work, service firms must depend by default on consumers' conscious rational responses. However, disconnects between what consumers consciously think and what they feel at deeper levels often lead to innovation failure.

In order to be effective in their innovation processes, service firms must reconnect the emotional, feeling dimension of customers' minds (right brain) with the perceiving, thinking (left brain) dimension of their minds to yield a holistic picture of the customer's mind. Some of the techniques designed to uncover latent customer needs (or "unarticulated needs") and tap into their unconscious minds include: ZMET (Zaltman Metaphor Elicitation Technique), Ethnographic Research, Story Telling and even Customer Hypnosis (McFarland, 2001; Zaltman, 2003; Reed, 2004).

Customer Ecosystem Mapping

Customer ecosystem mapping helps service firms build a strong foundation for future growth. Customer ecosystem mapping is about pursuing business opportunities well beyond the incremental adjacencies of traditional innovation. Customer ecosystem mapping provides a 360-degree picture of the entire customer value-chain and highlights a myriad of innovation opportunities. It also allows service firms to understand more deeply the needs of the different players at every point in the value-chain. Customer ecosystem mapping provides a much more thorough comprehension of the needs of the *customers* of one's customer and the *suppliers* of one's supplier.

Central to customer ecosystem mapping is a workshop where groups of constituents describe their activities around a specific element of the value-chain. Each workshop is done with a single constituent group. For instance, in the health care industry, patients, nurses, doctors and hospital administrators are more likely to be frank about their activities and the under-performing parts of a given process if they are in a group of their peers. Specially-trained facilitators should be used for these workshops. Facilitators should elicit frank responses from the participants, as well as help the participants uncover process deficiencies. They should encourage the participants to talk openly about their concerns as a whole—not just those pertaining to a given service or product.

The result of a customer ecosystem mapping workshop should be a highly-detailed activity map that an innovation development team can later use to prioritize opportunities. In other words, once the maps are created, the service firm (spon-soring the event) distills the creativity of the workshop into actionable learning. The first step is drawing some conclusions from all the information on the maps and adding rigor to the process of identifying deficiencies. Activities are grouped to help the service firm (sponsoring the event) locate where defi-ciencies identified by one constituency might overlap with those identified by another constituency. The insights surfaced in these workshops have surpassed any insights generated by ordinary focus groups or surveys of customers. In customer ecosystem mapping, participants are free to drive the conversa-tion rather than respond to questions designed by the sponsor-ing service firm.

The service firm sponsoring the event is originally involved only in one small part of the continuum. After the workshop is completed, however, the service firm sponsoring the event has now mapped out the entire value-chain of a given service for every constituency and is beginning to understand all of the dif-ferent services and variables that can affect the goal at the cen-ter of those charts: providing added value to the end customer.

One of the largest players in the renal care products market, completed more than 30 of these workshops and now has a much-improved understanding about the activities of renal care and all its constituents—patients, nurses, doctors, and hospital procurement officers. The company comprehends the range of unmet needs involved in the renal care-related activities (i.e., treatment of kidney disease) by mapping the activities of all the constituent groups, thus covering each stage of the treatment process. It translates those process gaps into opportunities for the company to intervene with products or services to improve the entire spectrum of care. The company moves beyond its traditional suite of existing products and services to find innovation opportunities in the whole process based on its deep knowledge of the renal care ecosystem (Corporate Strategy Board, 2001).

A well known commercial bank serving global institutional investors, has done several ecosystem mapping workshops to understand the top priorities of investment managers across the industry. Using this approach, the bank has developed an ecosystem map of end-to-end trade flows, which breaks the process into recognizable and commonly executed tasks. This ecosystem map is vital to the bank team's ability to understand the customer's trade ecosystem, because customers may characterize back and middle office processes differently.

During the workshops, the team poses its questions in such a way that the customer employee can provide an answer that furthers the team's understanding of the customer's needs and processes for trade transactions. Taking the information gathered during the workshops with the client, the bank builds its understanding of how that individual customer operates and where its priorities lie. The team reconstitutes the customer's trade ecosystem and compares that to its own benchmarks. This comparison allows the team to locate where the customer's trade ecosystem deviates from the standard and to identify opportunities for innovation (Corporate Strategy Board, 2001).

Customer Innovation

An organization's governance system to leverage all the intelligence acquired and intimacy developed around customers to innovate and deliver new, superior customer value.

In many firms, innovation is generally an unsystematic, collective process in which employees and managers participate in different interaction patterns at the formal and informal level. Service firms historically are not good at organizing the innovation process in a formalized and systematic way and learning from the process. The contemporary tendencies in the service sector, however, are towards a more systematic innovation process, often based on a structured and integrated process. Following the three basic approaches to developing a systematic customer innovation process are delineated:

(1) Process (managing a structured innovation process that covers idea generation, idea incubation, idea approval and idea implementation)
(2) Magnet Team (increasing innovation effectiveness with a concurrent forum for approval of ideas and a pipeline to manage future innovation)
(3) Product Forums (building innovation capital via knowledge exchange forums for product managers, marketing and sales)

Innovation Process

The notion of an "innovation process" varies across industries and enterprises. According to Robert Saco (Saco, 1997), when applied to services and knowledge work, the meaning of an "innovation process" may have a wide connotation. Some organizations have developed highly structured approaches to innovation, while others have promoted a work culture where innovation is a shared belief. A typical innovation process should address four different stages as follows: (a) idea generation, (b) idea incubation, (c) idea approval and (d) idea implementation.

Exploring ideas to fill identified gaps represents the first stage in the innovation process and requires a cross-functional effort to brainstorm ideas that will fill a certain need for a specific service segment. The second stage transforms the idea into a service concept that involves a combination of elements comprising the customer experience, including people, processes, products, and enabling technologies. The third stage requires the approval of the service concept by all the areas involved in the future development of the concept. The fourth stage is the implementation of the idea or new service concept and requires the development of a detailed measurement blueprint that includes cost estimates, timelines, and required control groups to accompany the service concept. Below is a more detailed description of each stage:

Idea Generation: Generating and Managing Idea Flow

This stage involves the generation, collection, evaluation and recommendation of new ideas for investment. To facilitate the process, top management should designate an Innovation Catalyst and establish an innovation council—called here a Magnet Team. The Innovation Catalyst's task should be to identify and establish expeditionary innovation projects strategically aligned with new business development goals. The Magnet Team's management responsibilities can range from screening ideas for new business opportunities to evaluating nascent projects for further development.

Many sources of ideas for new service concepts exist, and the six approaches for customer intelligence and customer intimacy are the major ones. As previously mentioned, ideas generated from "voice of the customer" surveys often result in incremental service improvements; truly innovative, new-to-world ideas often involve less input from customers and result from team ideation sessions. The Innovation Catalyst, working with the Magnet Team, should not only seek to align innovative solutions with current customer business line priorities, but also strive to achieve breakthrough innovations. Note that an Innovation Catalyst should possess the following characteristics: (a) have strong experience across service products and functions, (b) have respect within the organization for his/her achievements, (c) have the ability to manage interface across different groups, and (d) have solid facilitation skills.

Idea Incubation: Enriching and Structuring Ideas

Ideas grow in value as many connections are made between them. Therefore, whenever a product manager (or any member of the organization) devises a new idea, the Innovation Catalyst should connect this person with other co-workers within the organization to promote the connection of ideas. The Innovation Catalyst actually works as a "human portal" connecting and networking with people. This action will enrich and build intellectual capital around the original idea. The incubation process is conducted only for selected ideas and culminates in a final appraisal of the new service's or product's economic viability and market potential by the Magnet Team.

Idea Approval: Concurrent Approval and Cross-Functional Support

In this regard, the Magnet Team not only analyzes and approves ideas, but it also provides monetary resources and constructive criticism and exerts time pressure to guarantee the service's or product's continued development in an efficient manner. To coordinate the innovation process, the Innovation Catalyst (which is designated by top management) functions as an "orchestrator" to ensure the innovation process, is supported by key mechanisms (e.g., Magnet Team, customer intelligence, customer intimacy), and works efficiently in generating and cultivating ideas for new business development that are aligned with corporate goals.

Idea Implementation: Implementing and Executing Approved Ideas

The actual implementation of ideas is just as important as their generation. The team in charge of the implementation should develop a detailed measurement blueprint that includes cost estimates, timelines, and required control groups to accompany the design concept. Once the implementation team builds the operating framework for the experiment, it should test the design concept through rapid cycle experimentation. To temper the effects of any other variables other than the test idea, continuously repeating experiments and tests to further isolate the tested variable is prudent.

While research and development is a well-developed science for product firms, research and development in the services industry is still in its infancy. The lack of a precise "services" R&D methodology often leads services firms to invest in service initiatives without a clear understanding of their potential returns. Service firms can quickly leave themselves vulnerable to new-to-world service innovations brought to market by competitors, as service concepts quickly become "replicable" without a disciplined approach to research and development.

Sustainable services differentiation requires that service firms develop a disciplined R&D approach that allows them not only to consistently meet customer expectations, but also to develop distinctive approaches to service delivery that surpass customers' ever-rising expectations.

Magnet Team

The Magnet Team should be a "concurrent" forum for approval of ideas and a "magnet" to foster an innovation culture. For that reason, the Magnet Team should be a cross-functional team of managers with approval powers to review and approve ideas about new services or products. Following is a brief description of a typical Magnet Team.

- The Magnet Team consists of members representing product areas, legal, risk management, financial control, operations, technology and compliance.
- The Magnet Team can be comprised of people lower than the functional heads, but these representatives must be approved and empowered by their respective functional heads.
- Each member of the Magnet Team should designate a back-up to attend meetings in his/her absence. By being part of the Magnet Team, the members/back-ups agree to attend all meetings.
- Meetings are called by the Innovation Catalyst either on a pre-designated day of the week or on an ad-hoc basis.
- The Magnet Team is a horizontal organization with all members having equal authority.

Following are some measures used by the Magnet Team to evaluate the effectiveness of the innovation process:

- New service product revenues: Service products considered new for two years after they are launched
- Innovation index: New service product revenues divided by total revenues
- Cycle time: Period of days from the Magnet Team approval date to the date when revenues are enter into the books
- Ideas approved: Number of ideas or new service concepts approved by the Magnet Team
- Ideas rejected: Number of ideas or new service concepts rejected by the Magnet Team
- Conversion ratio: Number of approved ideas or new service concepts divided by the number of total ideas or new service concepts presented to the Magnet Team

In order to foster an innovative culture, the Magnet Team should apply the following four tactics:

1. Celebration of failures and near-failures: The Magnet Team should be proud of rejected and unsuccessful experiments. Such boasting of nearly-failed services or products helps foster an innovative and entrepreneurial culture.
2. Clear Goals: The Magnet Team should set a clear performance challenge—e.g., expect to generate 30% of its annual sales revenue from services or products introduced during the preceding 12-month period.
3. Information Sharing: The Magnet Team should promote a Service/Product Forum, with members from product, sales and marketing areas. The Forum should convene on a regular basis (e.g., quarterly, semi-annually) to allow participants to share information about innovation success stories, new research avenues and new applications.
4. Manage the Innovation Pipeline: The Magnet Team should manage the pipeline of ideas and new service concepts to control the incoming flow of revenues associated with innovation. The Magnet Team should keep the innovation

pipeline free-flowing and make the "efficient implementation" of innovations the status quo.

A large U.S. bank, in an effort to embed service development and innovation into the institution's retail business model and brand, established its own version of the Magnet Team. Called the "Innovation & Development Team," its objective is to develop a scientific method for "service" proposition research and development. The "Innovation and Development Team" runs real-time, service-based experiments aimed at creating both incremental improvements and differentiated new-to-world service delivery concepts. To test and operationalize innovative service ideas, the bank uses selectively chosen branches as active laboratories where the team conducts tests with real customers, measures results precisely, and identifies profitable innovations for broader rollout. The institution rigorously tracks performance data to gauge the relative success or failure of each experiment and its impact on customer satisfaction, revenue generation, and staff productivity (Thomke, 2003).

Product Forums

Building innovation capital via knowledge exchange forums for product managers, marketing and sales is vital in the effort to promote an innovative culture. Product managers should be encouraged to pass on lessons learned about innovation to other product managers or employees and help them with new projects. In that regard, communities of practice should be created, and these individuals should be the owners of the content. Following are key objectives to be accomplished by the Product Forums:

- Help the entire organization become more innovative.
- Provide a forum for collecting knowledge and intelligence about innovation success stories and enable successful transfer across different parts of the organization.
- Provide product managers with an in-depth diagnosis of where the organization is with regard to being "innovation prepared."

- Combine innovation "best practices" knowledge to provide management with an actionable set of recommendations that should lead to stronger innovative performance.

BENEFITS OF THE CI-3 FRAMEWORK

The CI-3 framework can help service organizations develop advanced customer knowledge management capabilities as well as create new sources of, and markets for, customer knowledge in the context of future customer value propositions and market developments. Doing so requires new competencies to select and motivate customers, integrate customer knowledge, disseminate knowledge within the organization, and act on that customer knowledge to translate it into new competitive offerings. The CI-3 framework emphasis is on assisting service firms to acquire skills to manage new sources of, and markets for, customer knowledge in the context of future customer innovation opportunities.

The CI-3 framework can also be used as an assessment tool to evaluate the following:

- Existing customer data gathering, analysis and application activities.
- A benchmark of one's activities vis-á-vis best and "next" practice.
- Opportunities for developing new customer, marketing and partner relationships to strengthen one's customer knowledge gathering and learning processes.
- Potential and means to implement explicit and implicit incentives for employees to share knowledge and efforts which stimulate collaborative creation of knowledge among employees.

If part of an integrated management system, the CI-3 framework will be a vital component of the overall development and refinement of the customer innovation opportunity.

The CI-3 framework helps service organizations in diagnosing innovation, identifying unmet market needs, and building tools that properly guide and support their own innovation programs. Collectively, the methods presented in the CI-3 framework can help service organizations build a robust innovation competence and leap beyond their competitors.

Innovation in services requires a selective and deliberate approach to making investments. Instead of investing everywhere and without aim, successful service firms develop customer intelligence to identify, and then spend aggressively on, the elements of innovation that will have the most impact on their service offerings and value proposition. Realizing that customer intelligence has value is the first step in acquiring and strengthening it. If doing a simple customer satisfaction survey is the only way to gain high-level attention and make the point, then by all means one should do it. Move quickly, however, to a more holistic approach, such as understanding the brand attributes that are most likely to influence behavior, identifying the key drivers of customer loyalty, and using touch-points data management to spot trends and find nuggets of information about customers.

The CI-3 framework help organizations continuously learn about their customers and markets through the linked processes of customer intelligence and customer intimacy, which follow the usual sequence of information-processing activities. Customer intelligence, as previously described, includes the collection and distribution of information about explicit and articulated needs of customers through traditional inquiry. Customer intimacy, however, is the collection and distribution of information about unspoken and unarticulated needs of customers through unconventional methods. Customer intimacy can be acquired by either getting customers more involved in the development process (i.e., customer blending), and/or by observing them more closely in their own environments (i.e., ethnographic research).

The installation of an innovation process permits service firms to not only test new-to-world ideas that otherwise might be deemed unrealistic, but also to reject those ideas that seemed beneficial in theory but were either too expensive or too little valued by customers. By applying a process to continuously generate, test and roll out new service concepts, a service firm may remain ahead of decaying value propositions and surpass ever-rising customer expectations. The installation of an innovation process requires a real commitment of physical resources, management time, and staff resources to successfully develop a services innovation lab. Especially for service industries, conducting experiments in a live setting inevitably leads to disruptions in the regular ways of "doing business" for customers and can result in customer confusion and potential dissatisfaction around newly-introduced processes and/or technologies.

Furthermore, service firms should not only source ideas from customers and employees across all lines of business, but they should also dedicate a separate team of creative thinkers to identify and conceptualize breakthrough, new-to-world innovations. Experiments should run no longer than 90 days (unless a particular process requires a longer testing period) to make room for new experiments that have been stored in the idea inventory. Capacity planning is important for service firms, and they should manage the capacity for experimentation, both within each test and across the testing sites, in order to capture feedback and learning effectively.

CITIGROUP: A LEGACY OF INNOVATION

Many of the companies that became part of Citigroup started in the 1800s. Schroders and the Farmers' Loan and Trust opened in 1818 and 1822, respectively. Travelers, Smith Barney, Bank Handlowy, Banamex, and Golden State Bancorp's predecessors were born late in the 19th century. The International Banking Corporation (IBC), Salomon Brothers, and The Associates sprang up in the early years of the 20th century.

Product innovation and distribution are hallmarks of Citigroup's legacy companies. Travelers (insurance business) pioneered auto, aircraft, group life, and "double indemnity" life insurance and was the first to insure American astronauts. The Associates (commercial banking business) originated loans for Model T Fords, the first mass-market automobile. A Golden State Bancorp predecessor made the first GI Bill home loan to a World War II veteran. EAB (retail banking) predecessors pioneered Saturday banking hours and were the first to offer junior savings accounts. Banamex introduced ATMs, savings accounts, and personal lines of credit in Mexico.

Citigroup has a history of innovation in expanding boundaries. Schroders was one of the first foreign banks in Japan and also one of the first banks to finance the building of railroads there. Bank Handlowy was one of the few banks to support trade with pre-Soviet Russia and Western Europe. National City Bank was the first nationally chartered American bank to open branches overseas and to run a foreign exchange department. The merger of Chas. D. Barney & Co.'s brokerage house with Edward B. Smith Co.'s underwriting business in 1938 created an early full-service investment firm. Citibank was the first U.S. bank to offer ATMs, travelers checks, mutual funds, and negotiable certificates of deposit. In addition, the merger of Travelers and Citicorp changed the landscape of the financial services industry forever. National City Bank helped finance the first transatlantic cable in 1866 and the expansion of U.S. railroads.

Citigroup eased trade, underwrote roadways, and electronically sped funds from one corner of the globe to another. Citigroup made the earliest foreign loans by financing railroads in Mexico, Central and South America, and Japan, and helped launch fleets of jets and supertankers. Citigroup lent its funds and expertise to communities large and small. In the 1970s, it helped resolve New York's near-bankruptcy, provided trade financing to Korea during its oil crisis, and acted as a lifeline to Indonesia during its debt crisis. Citigroup has a 200-year-old legacy of innovation and achievement. During those years, it has succeeded because it has taken the long view of the business. Citigroup expects to follow the same approach for centuries to come.

TAKE AWAY

1. Services represent about 80% of the U.S. gross domestic product and between 60% and 80% of the GDPs of the rest of the world's advanced economies.
2. Client participation (in various forms) in the production of the service may be the most basic characteristic of service activities, particularly knowledge-intensive ones.

3. Innovation in services is either a change in *products or services* (which service organizations offer) or a change in the *ways* in which products or services are created and delivered.
4. Innovation in services can be a new service product, a new procedure for producing or delivering the service, or the introduction of a new technology.
5. Innovations in service firms must be explained from the customer side (i.e., "pull oriented"). Innovation in services is customer oriented, but innovation is also often developed from ideas within the service firm (i.e., "push oriented").
6. New perspectives are urgently required to meet both the potential and need for sustainable innovation and profitable customer value creation. The CI-3 framework is composed of three CIs and nine approaches. The CI-3 Framework includes the following:
    i. Customer Intelligence
    ii. Customer Intimacy
    iii. Customer Innovation
7. The CI-3 framework can help service organizations develop advanced customer knowledge management capabilities as well as create new sources of, and markets for, customer knowledge in the context of future customer value propositions and market developments.

PROTECTING THE INNOVATION

JUSTIN SWINDELLS

Innovation requires an investment by the corporation with the expectation that such an investment will bear fruit and increase the value of the firm. In today's knowledge economy where information can be shared quickly and broadly, protecting these benefits is important to create a competitive advantage. This protection can be achieved through the use of the legal system (e.g., by registering patents, trademarks and copyrights) or by keeping the innovation a secret so that it cannot be duplicated (e.g., the formula for Coke). Often firms do not protect their intellectual property (gained through innovation) until they are assured of its commercial value. Today, where almost half the patents are being registered by foreign corporations, it is extremely important that the U.S. corporations also protect their investment in intellectual property.

INTRODUCTION

An innovation is practically worthless unless it is adequately protected, and all the time, effort, and resources put into developing and perfecting an innovation could be wasted by failing to protect the intellectual assets. The particular process or methodology, such as the scientific method advocated in this book, used to develop an innovation is important to one's ability to efficiently produce an innovation that will be successful in the business world. In the legal world, however, all manners of traps, landmines, and pitfalls await the naive innovator. One fact that makes this country a fertile place for innovation is the constitutional protections put in place by our Founding Fathers to encourage public disclosure of innovations in exchange for limited monopolies. The unique legal system of the United States, backed by the Constitution, offers powerful tools for protecting innovations—tools which some innovators may ignore at their own peril.

The main paradigm shift in the innovator's thinking regarding protection should be simply this: move from not thinking about protecting innovations to thinking about protecting innovations. This paradigm shift is actually facilitated by the change in the innovation paradigm from pure art to science. Thinking about innovation as a scientific process leads one naturally toward efforts to protect the innovation. Although somewhat counter-intuitive, protection of innovation actually fosters and promotes more innovation. This paradigm shift from art to science is therefore a self-fulfilling prophesy, with greater innovation protections acting as a feedback loop to accelerate the pace of new innovation. In other words, the paradigm shift encourages increased innovation, which in turn leads to increased innovation protection, and that in turn engenders more innovation.

The innovations which are able to be protected cover the gamut. A starting point, as Congress so aptly stated, is that "everything under the sun that is made by man" is able to be protected. A few obvious exceptions exist, such as laws of nature, mental processes, mathematical formulae, fundamental truths, and even perpetual motion machines (until someone builds one that really works). Even mere improvements over prior innovations can be protected, and indeed, most innovations fall into this category. The question should not be whether an innovation can be protected, because all true innovations are, but rather how it should be protected. For the answer to this question, a variety of options are available, including patents, trademarks, copyrights, and trade secrets. Once the innovation is protected, exploiting an intellectual property asset can happen in various ways, principally licensing and litigation.

The most important concepts to keep in mind are (1) the sky's the limit on what can be protected, and (2) engaging in innovation is a pointless and wasted effort unless adequate steps are taken to protect it. The paradigm shift in innovation does not change these concepts. Many innovators tend to focus naturally on what they are best at—innovating—but do not recognize the perils of failing to protect their innovations. Imagine the frustration of developing (at great effort and expense) an innovation and commercializing it at a price that recaptures development costs, only to later discover a copyist duplicating the innovation with minimal effort and cost and selling a knock-off of the innovation at a lower price. Without protection, the innovator is powerless against the copyist.

Some innovators feel a tension between spending the resources to protect an innovation and resolving an uncertainty that the innovation will be successful in the marketplace. In other words, the innovator is willing to spare no expense developing and perfecting an innovation that is intended to be successful in the marketplace, yet a lack of confidence may convince the innovator to take the risk of not adequately protecting that innovation. Even minimum protection is better than no protection at all. It makes little sense to expend so much energy producing an innovation only to express doubts later about whether it is worth protecting. More to the point, the patent laws allow an innovator a grace period of one year to test out the marketability of the innovation before spending any resources protecting it. Thinking of innovation as a scientific process rather than an art form will discourage irrational decisions that lead to penny wisdom and pound foolishness.

TOOLS TO PROTECT THE INNOVATION

Four basic tools may be used for protecting an innovation: patents, trademarks, copyrights, and trade secrets. The first three tools reward the innovator with limited monopolies in exchange for disclosing ideas to the public. The fourth, trade secrets, are just that—secrets—and are effective for as long as the innovator can keep the innovation secret. A famous example is the recipe for the Coke brand soda. Maintaining an innovation as a trade secret is generally more difficult than to disclose it to the public in exchange for a limited monopoly.

Congress created a set of intellectual property laws intended to encourage innovators to disclose their innovations to the public, so that the knowledge base of ideas would increase, thereby spawning more innovation. If innovations were not quickly revealed to the public, innovation would be stifled, because people would be reinventing the wheel over and over again without the benefit of everyone else's prior ideas and efforts. Trade secrets are not unworthy, however. On the contrary, particular innovations are very appropriate candidates for trade secret protection. On the whole, however, most innovations benefit from the bargain created by Congress—public disclosure in exchange for a limited monopoly.

The phrase "limited monopoly" is controversial in some quarters, because it is perceived to have a negative connotation. Without using lots of qualifying legal jargon, "limited monopoly" in essence means the innovator is given power to exclude others from infringing upon the space he claims as his intellectual property.

Take a patent as an example. A patent is basically a bundle of rights that is akin to a deed to a real estate parcel. A deed describes in words the boundaries of the property to which the owner claims to own in the real estate landscape. In the same way, the "claims" of a patent are words that describe the boundaries of what the innovator claims to own in the intellectual property landscape. However, whereas deeds describe boundaries that define what owners can do within those boundaries (build a house, lease the land, etc.), patent claims describe boundaries that define what infringers cannot do. In other words, patent claims are like "Keep Out" signs, granting patent owners the right to keep others away from their claimed intellectual property space.

This "right to exclude" can be a potent weapon against would-be competitors. A patent owner does not actually have to make the patented product or use the patented process. This means that an innovator can develop an innovation and license someone else to make and sell it. In exchange, the innovator promises not to sue the licensees, and in exchange for that promise, the licensees promise to pay the patent owner royalties (usually money), all without the patent owner's lifting a finger to commercialize the innovation.

The right to exclude can also be a potent weapon against actual competitors who exploit a patented innovation without the patent owner's permission. Patent owners can enforce their exclusive rights through litigation against such competitors and stop them from making, using, and selling the patented innovation. In some cases, such action can put a competitor out of business.

If innovators find themselves on the receiving end of such a threat, imagine the enormous leverage created by having a portfolio of intellectual property assets to throw back at the patent owner. Thus, patents and other tools can be used for purposes other than merely protecting one's own innovations. They can also be used to create zones of protection within a competitor's space, thereby at least encouraging the competitor to think twice before asserting some of its own intellectual property assets and to consider the possibility that it may be unlawfully encroaching upon someone else's territory.

These strategies and many others are interesting and effective, but they assume that one has assets to protect in the first instance. Without assets, the only strategy to worry about is how long it will be before someone tries to shut down the innovator's business. The answer to that question is "about as soon as the business becomes successful." Successes breeds attention, and though imitation may be the sincerest form of flattery to some people, why even allow it in the first place? Naked innovators become most vulnerable when they achieve success, often at a point in time too late to seek protection.

How many innovations are patented? So far, over seven million patents have been granted since July 31, 1790, and the number is increasing. In 1963, the United States Patent and Trademark Office reported that about 91,000 patent applications were filed. In 2003, about 366,000 patent applications were filed—a four-fold increase. From 1993 to 2003 alone, the number of patent applications filed almost doubled. Countless other potential patent applications were never filed, because the innovator either failed to recognize the value of patent protection or consciously determined that the risks of having no patent protection were not outweighed by the benefits of having adequate protection. These types of innovators, no matter how brilliant their innovation or the methodology they adopted to create the innovation, are unwisely exposing their innovation to an unnecessary risk of failure.

Part of the reason for the sharp increase in patent applications is that foreign companies are increasingly taking advantage of the attractive protections offered by U.S. Patent laws. In 1963, only 18% of patents were issued to foreign residents, but in 2003, the number jumped to 47%. Thus, nearly *half* of all patents today are issued to foreign residents. If any remaining doubt exists about the wisdom of seeking the protection offered by U.S. Patent laws, one simply needs to consider that half of the patentable innovation today is foreign-based.

Unlike real estate deeds, patents have fixed expiration dates. The right to exclude does not last forever (hence the term "limited" monopoly). Today, the term of a patent is fixed to no more than twenty years from the earliest effective filing date of the patent.

Patents used to theoretically have an indefinite term using a now discredited strategy infamously known as "submarining." A typical submarine patent is one that was filed decades ago as a "blue sky" patent application, but kept alive through a chain of related patent applications. As new technology emerged over the subsequent decades, the patent owner wrote claims to cover the new technology, which was usually based on a tenuous interpretation of the original blue sky patent application. Innovation is everywhere—even in the legal field. In the legal field, however, "innovation" strategies (such as submarining) usually only work once. The playing field must be fair, and with few exceptions, it is.

Still, even without submarining, twenty years from filing is a relatively long time, particularly considering the present rate of technology obsolescence. Provisional protection can begin as early as eighteen months after the filing date of a patent application, yielding as much as eighteen-and-one-half years of protection. That period is a long time to exclude others from implementing the protected innovation or to collect money for permitting others to implement it. It is also a long time to allow a competitor to appropriate an unprotected innovation with impunity.

A special kind of patent, called a design patent, protects the ornamental design of a manufactured article. It does not protect the underlying article itself—for that one needs a patent. As stated by a federal court of appeals, design patents are intended to eliminate the "unsightly repulsiveness that characterizes many machines and mechanical devices which have a tendency to depress rather than excite the aesthetic sense."

Value exists not only in the article itself, but also in the aesthetic attractiveness of the article. The success of Apple's iPod players is due in no small part to its aesthetic design. Design patents are a reminder that innovation does not concern technological advances alone, but also the way the technology is packaged and presented to the consumer. Just as the innovation process is not pure science or pure art, so too intellectual property protection does not concern purely science or purely art.

Patents protect the details of the innovation itself, while trademarks protect the brand that is used to market the innovation. The importance of branding an innovation is obvious. The difficulty is coming up with an effective brand, aided with the principles set forth in this book. Once that brand is selected, the proper protection device is a trademark (or a service mark).

A trademark is basically a brand name that is used to distinguish the goods of one person from those of someone else and to indicate the source of the goods. A service mark is essentially the same thing, except it applies to services instead of goods. A trademark or service mark does not protect the innovation itself (unless of course the innovation is the brand), but rather it simply protects the *brand* associated with the innovation. To protect the innovation itself, one should seek patent or copyright protection, as appropriate.

Trademarks are registered by the U.S. Patent and Trademark Office, and owners of federally registered trademarks enjoy the exclusive right to use the trademark in the territory in which the mark is registered, which is presumptively the entire United States. The marks are registered for fixed terms that can be renewed for a fee as long as they continue to be used properly. Thus, trademark protection can last much longer than patent protection.

Thinking of numerous trademarks that have existed for decades and that continue to remain in force today is not difficult. Even though the innovation designated by a brand name may no longer be "protectable" under the patent laws, the brand associated with that innovation can continue to be a market differentiator long after the patent protection has lapsed. Innovators would do well to create and establish a strong brand during the patent term so that purchasers continue to associate the product or service with the brand even after it no longer enjoys patent protection. Good branding yields "goodwill," which can have tremendous value. Many brands today have significant goodwill and are valued at a considerable amount of money for what amounts to nothing more than the right to use a few letters or symbols to designate a product or service.

Trademark rights, like patent rights, can be licensed. Trademark owners should police their brand to ensure that it is being used properly and not in an improper or unflattering manner. The cost for failing to police a brand adequately is dedication of the brand to the public or the inability to collect money from the trademark usurper. The refrigerator, cellophane, aspirin, the thermos, and nylon are a few examples of once-famous trademarks now dedicated to the public because of inadequate policing.

Certain innovations, like music, photos, movies, software, and literature, enjoy automatic copyright protection if they are original works. Copyrights can be registered, but protection is automatic, unlike machines or processes seeking patent protection. Because artists enjoy automatic protection for their original works, the paradigm for artist innovators tends to be creation-centered, not protection-centered. On the other hand, patent innovators must affirmatively seek out protection for their ideas; otherwise the default is no protection.

Grossly oversimplifying the contrast between copyrights (art) and patents (science), artist innovators do not need to ask the question, "How do I protect my original work," because protection attaches to the work as soon as it is reduced to a tangible medium. On the other hand, patent innovators must ask the question or risk having no protection at all. This contrast between copyrights (art) and patents (science) can be directly analogized to the paradigm shift from innovation as an art to innovation as a science. That is, as the innovation process shifts from pure art to science, the innovator's frame of mind naturally tends to become more protection-centric, partly as a self-preservation response. Approaching innovation as a scientific process, the innovator will logically seek to protect the ideas embodied in the innovation.

The tension between the arts and the sciences can be understood as a problem of ownership. Artist innovators sign their painting, credit themselves in a film or song, or put their names on a book or a musical composition, and that signature or accreditation constitutes an imprimatur of ownership. The tangible media (canvas, film, CD, paper) on which the idea is expressed is itself not patentable, so science innovators cannot rely on a mere claim of authorship to protect the *ideas* expressed on the paper. A claim of ownership may establish origination of the idea, but it offers no empowerment to stop someone else from taking the idea and using it. A copyright only protects the *expression* of an idea, not the idea itself. Ownership of a new and useful *idea* (as opposed to its expression) requires the innovator to place a stake in the ground declaring that idea as his own.

This burden on the innovator to make affirmative efforts to claim an idea exploits the self-preservation instincts of the innovator. This self-preservation instinct will eventually lead the innovator to think about protecting the idea as the innovation thought processes are exhausted. Changing the framework for innovation thinking from pure art to science makes sense. The science-oriented framework presents a self-sustaining, iterative process that yields more innovations, which in turn causes increased efforts to protect the innovations, yielding still more innovations as rewards from the protection efforts are realized. More innovation yields benefits for all, because the progress of useful ideas is advanced by knowledge awareness and sharing, allowing new innovators to build upon the previous efforts of others. This policy of encouraging prompt disclosure of new ideas is expressed in Article I of the

Constitution, which empowers Congress to "promote the progress of science and useful arts." The science of innovation advances this fundamental policy.

USING THE RIGHT TOOL EFFECTIVELY

Protecting the innovation is half the battle. Indeed, some protection is better than none at all. However, good protection is a matter of selecting the proper tool and applying the tool effectively according to a deliberate strategy. Experienced patent lawyers should be consulted as part of any effort to protect intellectual property assets, just as medical doctors should be consulted to diagnose and address medical problems. The do-it-yourself approach rarely works and can produce more legal problems.

How one protects intellectual property adequately is as much a science as is the process of innovating intellectual property. A significant amount of thought should be invested in developing a strategy for protecting intellectual property assets. For example, every year thousands of laborers spend countless hours planning their vacations instead of spending even one hour to plan an investment strategy for retirement. The allure of maximizing enjoyment from the fruit of labor colors the vision one needs to protect that fruit. In a similar way, innovators can fall victim to this dangerous tunnel vision.

On the other hand, what rewards await the innovator who thoughtfully puts in place a strategy to protect his intellectual assets? Numerous examples of effective strategies abound, but for every effective strategy that is carried out, a competitor is wincing on the receiving end, either from the effects of its own inadequate protection, or from an effective enforcement strategy waged against it.

SIGNIFICANT PATENT DISPUTES

Take the fairly recent example involving a patent dispute between privately held patent holding company, NTP, Inc., and the Canadian maker of the wireless BlackBerry® devices,

Research in Motion (RIM). After receiving mixed results in the federal courts that did not entirely resolve the dispute, NTP and RIM agreed to a settlement in which RIM paid NTP $450 million. The obvious risk for NTP was zero recovery. But a significant risk for RIM, which exists in nearly every patent dispute, was the prospect of an injunction that would have prevented RIM from selling its popular wireless BlackBerry® devices for as long as NTP's patent was enforceable.

The case between NTP and RIM is an example of how protection works. Stories abound regarding how an enterprise designed and built a better mousetrap, successfully implemented an effective marketing campaign to promote the mousetrap, and made a fortune. But numerous enterprises are quietly staking out zones of protection in the intellectual property landscape and reaping sometimes life-preserving financial rewards from those efforts. Foreign enterprises in particular have been taking notice for many years, owning nearly half of all patents issued today by the Patent Office, and that number continues to rise.

Intellectual property assets can rescue an enterprise from the brink of bankruptcy or condemn an exposed and unprotected enterprise to extinction. When used as a shield, these assets provide leverage or, at a minimum, an uneasy truce among competitors; when used as a sword, they can provide a significant source of revenue and enhance or even ensnare a market share. Small enterprises can use intellectual property to conquer Goliath enterprises in the marketplace, which would otherwise crush the small enterprise through sheer market power.

Immersion, a relatively small California company with fewer than 150 employees but with a very large patent portfolio consisting of over 270 patents and 280 pending patent applications, recently won a $90 million judgment against Sony for its infringement of a patent that covered Sony's PlayStation systems and games. The court also ordered Sony to stop making, using, selling, and importing the infringing systems and games permanently. Enforcement of that order has been deferred pending Sony's appeal, but permanent injunctions are a terrifying prospect for any company. A company faced with a permanent injunction faces several stark choices: exit that market altogether, obtain a license if the patent owner is willing to grant one and pay royalties, or redesign the infringing product, sometimes in a way that is inferior to the patented way.

Toshiba may soon be in the same boat as Sony. A jury recently awarded Lexar Media what is believed to be the largest intellectual property verdict in California history, and the third largest such verdict in U.S. history, against Toshiba in the amount of $465 million. Lexar Media's intellectual property portfolio includes over 80 patents, as well as various trade secrets that Lexar Media accused Toshiba of stealing. The award reflects in part the jury's evaluation of the value of Lexar Media's intellectual property. The case is still ongoing, and the award may be overturned or reduced. These cases aside, intellectual property assets can have tremendous value and wield powerful leverage over a would-be trespasser.

Japanese companies are hardly the only targets of innovators who hold patents and trade secrets. Two computer chip makers dueled over allegations of patent infringement and trade secret theft. Before trial, China-based SMIC paid $175 million to Taiwan-based TSMC in exchange for a cross-license between the two companies, in which each company promised not to sue each other for patent infringement. Whether money changes hands in consideration for a cross-license generally depends on whether a disparity exists in the valuations of the licensing parties' respective patent portfolios. The weaker portfolio holder usually has to pay to get the cross-license.

Some companies are forced out of markets altogether. In 1991, Kodak paid Polaroid a landmark $925 million to settle claims that Kodak infringed Polaroid's patent covering an instant photography process. Part of the settlement agreement required Kodak to exit the instant photography market then dominated by Polaroid. As the patent owner, Polaroid was not obligated to allow Kodak to compete in its patented space and could keep competitors at bay and out of its space for the duration of its patent.

Innovators with patents can also target an entire industry—not just one or two companies. Forgent began a licensing program three years ago that has so far netted over $100 million in licensing revenue from 37 American, European, and Japanese companies. What did Forgent wield that brought in this revenue? A solitary patent that was asserted by the law firm of Jenkens & Gilchrist, that had acted as Forgent's long-time general corporate counsel and that structured and developed a licensing program for Forgent around this solitary patent. Jenkens & Gilchrist then asserted the patent successfully for Forgent against some of the largest global consumer electronic companies.

Honeywell won a $96 million judgment against Minolta for Honeywell's claim of infringement of its patented auto-focus technology, and then proceeded to collect another $300 million in licensing revenue from other major camera makers. Honeywell did not use its patented auto-focus technology in its own products, but nevertheless successfully commercialized the technology by enforcing its patent rights. Innovators should take note that even if an innovation does not fit within a current or future business plan, the innovation may still be worth protecting, because others may someday decide to implement it, as in the case of Honeywell's auto-focus technology.

Rambus, a technology licensing company, sued Germany's Infineon Technologies for patent infringement of Rambus' patented memory technology. Rambus and Infineon settled their legal disputes in March of 2005, and Infineon agreed to pay a quarterly license fee capped at $150 million. After securing that deal, Rambus has reportedly set its sights on other memory vendors.

Successful companies attract attention because of their deep pockets, and sometimes in ways that expose vulnerabilities in their zones of intellectual property protection. Some companies innovate so rapidly, sometimes paying little heed to what intellectual properties they may be trespassing upon, that they get blindsided by a patent that seemingly emerges from nowhere. Valuable company resources normally invested in innovation and growth must now be diverted and expended to fend off the legal claims. For the patent owner, the patent may present an opportunity to capitalize on someone else's market share or to realize additional revenues, sometimes without even building any factories.

Microsoft is a frequent target of patent infringement claims. In the last decade or so, Microsoft was on the receiving end of several multi-million-dollar verdicts and large settlement disbursements. For example, a jury awarded Stac

Electronics a verdict of $120 million against Microsoft for Stac's claim that its patents covered some of Microsoft's legacy operating systems. The parties later reached a settlement agreement in which Microsoft agreed to pay Stac $43 million and to buy a $40 million stake in Stac.

Presently Microsoft is embroiled in a patent case brought by Eolas Technologies and the University of California, which jointly own a patent that apparently covers certain technology in Microsoft's Internet browsers. Eolas and the University were awarded a $521 million verdict, which has since been overturned and sent back to the trial court where the legal battle continues. In March of 2005, Microsoft agreed to pay $60 million to Burst.com to settle a patent infringement and antitrust dispute and to obtain a license to Burst.com's patent portfolio. In 2004, Microsoft settled with Sun for $1.6 billion over an intellectual property dispute, resulting in a technology-sharing arrangement between the two rivals. All these cases demonstrate, regardless of which company may be legally correct, the enormous leverage intellectual property can wield.

THE INCREDIBLE VALUE OF INTELLECTUAL PROPERTY PROTECTION

Intellectual property can also drive enterprises into bankruptcy or produce significant value for creditors in bankruptcy proceedings. In 1997, Paragon Trade Brands, a baby diaper manufacturer, filed for bankruptcy after Procter & Gamble was awarded damages of $178 million for its patent infringement claims against Paragon. When disk-maker Orca Technology filed for bankruptcy with $2.65 million in liabilities, the bankruptcy trustee auctioned off three patents owned by Orca. Bidding started at $225,000 but ended at a staggering $3.65 million, which was enough to repay all of Orca's creditors in full plus interest and leave a surplus of some $300,000, a veritable nonpareil in bankruptcy cases.

Another consequence of non-protection is simply lost opportunities. A classic example is the lost opportunity that befell Xerox in the 1970s. Innovators at Xerox's research center in Palo Alto developed a mouse-driven personal computer with a graphical operating system complete with icons and windows, a laser printer, object-oriented programming, and other fundamental technologies that are now ubiquitous. Xerox did not consider these innovations to be part of its core business and failed to patent them. The rest is history.

The Xerox example highlights the importance of protecting not only core technologies but also auxiliary technologies, like Honeywell's auto-focus or Xerox's personal computer technologies. Protection produces assets, assets provide leverage, and leverage strengthens bargaining positions. By protecting auxiliary technologies, the innovator creates additional leverage by staking claims in other areas in which it may not intend to compete. This type of auxiliary protection can be used offensively or defensively. Used offensively, the innovator can target a trespasser in the auxiliary space and force the trespasser to pay royalties. If the trespasser has no protection covering the innovator's commercialized technologies, the innovator's leverage over the trespasser is practically unbounded. Used defensively, the innovator can keep a competitor at bay that may be using the innovator's protected auxiliary technology even though that competitor has protection that covers the innovator's core technology. In other words, if the competitor tries to lob a missile, the innovator has a missile to lob back. Auxiliary protection is overlooked by innovators who think only about protecting their core innovations, and they do so at their peril.

Eventually, nearly all successful innovators will find themselves at the bargaining table, because success draws attention, and it is alluring to ride the coattails of those who have already forged the way. Whoever is sitting on the other side of the table, be it a close competitor, an obscure company, a holding company that makes and sells nothing, or a Goliath with thousands of patents in its portfolio, has but one relevant characteristic—an innovator with an intellectual property asset upon which it seeks to capitalize. What will be in the "victim" innovator's portfolio when forced to deal with this asset? This rival innovator sitting at the other end of the table is not going away. A victim innovator cannot simply devise a business strategy (although a business solution should be considered) and use its business expertise to beat this rival innovator out of the marketplace. Patents are great equalizers in that respect.

Long before the victim innovator is hauled to the bargaining table, it should have protected its core and auxiliary innovations through a well-developed strategy (as outlined in Chapter 14) that best fits its situation. A strategy that works for one innovator probably will not work for another. Developing an effective strategy is a complex undertaking involving consideration of business relationships, existing and prior technology, the relevant intellectual property landscape, cost, timing, existing intellectual property portfolios, and the competitor's intellectual property portfolios, to name a few. Still, innovators would do well to consider certain aspects in developing a strategy for protecting their innovations:

- Maintain confidential invention repositories that include proof of the innovations and the dates they were conceived.
- Protect the branding of the innovations by filing for trademark or service mark registration.
- Use the mark consistently, police it, and create a strong brand so that long after the patents expire, consumers continue to associate with the brand.
- Pay the renewal fees on time to keep patents and trademarks from lapsing.
- Keep chains of patent applications pending so that claims can be written to cover future products and methods.
- If auxiliary innovations are developed, protect them with patent applications.
- If protection is not sought on an auxiliary innovation and the innovation is not a trade secret, defensively publish details of the innovation to preclude another would-be innovator from later seeking patent protection on it.
- Survey the intellectual property landscape, especially in areas surrounding core innovations, to locate gaps in protection.
- Maintain and update charts that map out the zones of intellectual property for which protection has been sought.

- In consultation with a patent attorney, weigh the benefits and risks of periodically monitoring competitors' patent portfolios to identify newly issued patents and published patent applications.
- Purchase intellectual property to quickly firm up weaknesses in the portfolio or to establish a zone of protection in an auxiliary space.
- Do not toss out alternative designs that are not commercially implemented; these alternative designs could be useful for avoiding patents at a future date.
- Monitor licensing activity and patent litigation relating to core innovations.

PATENTING THE INNOVATION

For most purposes, a patent is the tool of choice for protecting an innovation. Legally, innovation includes two components: invention and reduction to practice. Invention is the formation in the mind of the idea with enough detail to build it. Reduction to practice is basically the construction or implementation of the idea. Actual reduction to practice is not required before filing a patent application. One only needs to have invented something patentable in order to file for protection. Building a working model is not necessary.

Ideas are inherently intangible. Therefore, a critical part of the patenting process is to document these ideas on paper. Documenting the early moments of the innovation process is important because, unlike every other patent system in the world, our unique patent system rewards the first to invent, not the first person who rushes to the Patent Office with a filing. One way to attack the validity of a patent is to demonstrate that someone else disclosed its teachings earlier, so if the inventor can prove an even earlier invention date, the patent may still be valid. Without written documentation, proving the moment of invention is more difficult, relying upon the inventor's historical recollection of how the invention unfolded. The written documentation should include text and drawings describing and illustrating the invention in sufficient detail to show a firm grasp of the idea.

The Patent and Trademark Office has a program called the "Document Disclosure Program," which allows inventors to send their documented inventions to the Patent and Trademark Office as evidence of the date of conception of an invention. The document must be signed by at least one inventor, and must not contain any non-paper attachments. The Patent and Trademark Office will hold this document in confidence for two years before destroying it. Before it is destroyed, the inventor can, for a fee, request a copy of the document. The copy should be preserved in confidence for safekeeping as evidence of the date of invention.

Ideally, before any public disclosure of the innovation, a patent application should be filed describing the innovation and claiming key aspects of it. New patent applications are held in secrecy at the Patent Office for 18 months, after which they are published to the world. If foreign protection is to be sought, it is especially important to file for a patent application before any public disclosure, because in most foreign countries, any public disclosure bars the filing of a patent application. Public disclosure does not bar the filling of a patent application in the U.S., however.

As mentioned earlier, the "do-it-yourself" approach to protecting innovations is as good an idea as self-treating an injury that otherwise would require medical expertise. Patent attorneys are lawyers with engineering or science degrees who have taken and passed a vigorous, specialized examination offered only to those with engineering or science degrees by the Patent Office. Patent work is one of the few areas of law that is particularly specialized, with its own bar examination, an appellate court that exclusively handles patent appeals, and rules and laws that govern conduct before the Patent and Trademark Office and the process of seeking and obtaining a patent.

Important aspects of the patenting process—drafting the claims that appear as numbered paragraphs at the end of each patent and convincing a patent examiner to issue a patent—require skillful use of words and knowledge of patent law that takes years to refine. (After all, claims are just a sequence of words put together in a careful way.) How much "real estate" is staked out on the intellectual property landscape depends largely on how well written the claims are. Well-written claims can also quickly convince trespassers, who might otherwise scoff at license overtures (thus forcing a decision to litigate or abandon the effort), to take a license or exit the space covered by the claims.

The process of applying for and obtaining a patent is called "prosecution," which unfortunately also refers to the government lawyers who handle criminal cases, but otherwise has nothing to do with criminal law. Patent prosecution strategies should be tailored for each innovator's situation, but a few common strategies are worth mentioning.

One strategy, particularly if cost is a consideration, is to file something called a provisional patent application. These applications are not examined by a patent examiner, so no patent can be issued, but these applications can be converted within one year to a normal patent application that is examined. They do not require any claims, one of the most expensive parts of a patent application, and their filing fee is a fraction of that of a normal application. Provisional patent applications are usually filed if there is a pressing need to get something on file quickly, if cost is a concern, or if there is uncertainty about the commercial viability of the innovation. If provisional patent applications are not converted to a normal patent application within one year, however, they go abandoned forever and cannot be revived.

Another strategy is to ask that the examination of a patent application be expedited, particularly if a competitor has a product on the market that is believed to be covered by claims of the patent application. Early issuance permits the patent owner to seek an injunction and money damages, which can begin to accrue as soon as the patent application publishes. The only catch is that the claims of the issued patent must be almost the same as the claims as they were initially published, meaning that the claims cannot be drastically changed from publication to issuance. Other reasons exist for seeking rapid issuance of a patent, which under normal circumstances usually takes about three years.

Early publication of a patent application can be requested to begin the money damages clock earlier. Early publication also makes the published patent application available to be used to knock out patent applications filed later, so this strategy can be used offensively and defensively.

If early publication is going to be requested, inventors would do well to conduct a thorough search for potentially relevant historical items. Another common misconception about patent prosecution is that the inventor does not have to conduct an independent search for historical items that may be close to the claims set forth in the patent application. These historical items are called "prior art," which is patent-speak for published documents and public devices that predate the invention and which may be relevant to the patentability of the invention.

An inventor has an important obligation to disclose all prior art known to the inventor, but has no obligation to conduct a search for prior art. The Patent Office will do its own search, which is part of the application fee. For early publication, however, it is important that the claims as originally filed remain unchanged through issuance; otherwise if they are changed significantly, money damages will not begin to accrue on the date of publication. Conducting a thorough prior art search before filing the claims will ensure that at least some of them are likely to be allowed by the patent examiner without requirement of any changes. If the examiner believes that a prior art item reads on a claim, the examiner will reject the claim, and in order to obtain its allowance, the inventor must change the claim until the examiner is convinced that it is patentable over the prior art item.

A frequently used strategy is to file a chain of applications all related to the original patent application to keep prosecution alive even after the original patent application issues. If an innovator expects a competitor to launch an infringing product, but does not yet know the details of that product, the innovator can keep the prosecution alive by writing claims that cover that product as soon as it is launched. Expedited examination and other strategies can also be employed to lead to a quick issuance.

The following scenario happens quite frequently with savvy intellectual property protectionists and illustrates a real benefit of filing additional related patent applications in order to keep the prosecution alive long after the original patent application issues. For example, a company may be approached for a license because a patent owner believes that the company is making products that infringe on the patent owner's patent. But the patent owner is unaware that the company actually has a patent application pending whose claims can be rewritten to cover a product made by the patent owner. Suddenly, the company has an asset to leverage against the patent owner, perhaps not for a few months down the road, but the patent owner must now live with the uncertainty that a patent may be issued covering products made by the patent owner. Now, instead of one-way licensing discussions, talks may turn to other solutions, such as a cross-license or a "walk-away."

Theoretically, because submarining is no longer permitted, a chain of related applications can be kept alive for 20 years after the filing date of the original patent application. The courts, however, are beginning to frown upon these extended chains of applications and are finding ways to prevent enforcement of certain patents that are issued long after the original patent application was filed through a series of related applications. One way to avoid this problem is to spawn numerous patent applications all related to the original patent application at the same time, by filing so-called "divisional" applications. These divisional applications will be examined in parallel (at least theoretically; in practice they will likely be examined in seriatim), and can be issued all within a few months of each other. "Mining early and mining deep" is the safest bet.

File a "jumbo" patent application disclosing numerous related inventions in one patent application. Multiple inventions can be set forth in a patent application, but only one invention can be claimed per patent application. An advantage to a jumbo patent application is that not only can each individual invention be claimed, but also the *combination* of different inventions can be claimed. If five different inventions are disclosed, then 120 different claimable combinations potentially exist that can be mined from a single patent application. After all, most patents are simply combinations of old ideas put together in a new way. To claim a different invention, the patent lawyer must file another patent application based on the jumbo patent application.

Many patent lawyers have discussions with inventors about the following misconception about patents. Consider this over-simplified example. Suppose a claim covers an alarm clock that triggers a metronome at a predetermined time. Alarm clocks and metronomes have been known for a long time. However, if no one ever thought to combine an alarm clock and a metronome before, that combination very well may be patentable. Recall that the starting point is that anything under the sun that is made by man is patentable. The sky is the limit, and when an inventor believes that an idea cannot be patented, a skilled patent lawyer can usually find patentable subject matter lurking in the idea itself or in combination with something else. Sometimes the subject matter is simply too narrow to be worth protecting, but the only limit to what is patentable is imagination itself.

Another strategy works only within two years after a patent is issued. A patent rule allows claims of an issued patent to be *broadened* if the inventor declares that a mistake was made without deception, but only within two years after the patent issues. After two years, the claims cannot be broadened. This procedure to reissue a patent can be exploited in some situations to change the claims to cover a competitor's product that otherwise just barely avoids the issued claims. Reissues are time-consuming, and they require the patent owner to surrender the patent while the patent is being examined again, meaning that the patent cannot be wielded against anyone during the period of reissue.

These strategies are but a few of many in an arsenal that a patent lawyer can dispatch to ensure that an innovation is protected as broadly and deeply as possible. One thing to keep in mind is that laying the foundation for intellectual property protection early is crucial, even though the benefits will probably not be realized until years later. By the same token, it is never too late to implement a protection strategy. Unfortunately, many companies only begin to implement a protection strategy until after they have been burned by the effects of having none. Still, intellectual property is probably one of the most overlooked and underappreciated strategic assets in the business world.

SUMMARY

Changing the paradigm of innovation from pure art to science naturally leads innovators to think about intellectual property protection primarily as a self-preservation response. The paradigm shift in how innovation is conceived leads to an ever-increasing rate of innovation. The legal system is structured in roughly an analogous way. Artistic innovators (art) enjoy automatic protection for their mental output, but technological innovators (science) must take affirmative steps to protect their mental output. The U.S. Patent system was founded primarily upon the correct belief that offering protection for innovators who promptly and publicly disclose their innovations in exchange for a limited monopoly will advance the progress of science and useful arts. Shifting the paradigm for innovation only further validates this vital policy.

The framework for innovative thinking set forth herein will naturally cause innovators who follow this framework to think about protection, and that is nine-tenths of the battle won. What plagues some innovators today is their inability to recognize protection as even an option. Those who change their thinking and approach innovation as a scientific method will become mass producers of innovation, because protection is a necessary part of any feedback loop to produce more innovation. Take protection out of the loop, and innovation is reduced to a series of discrete, inefficient efforts. Some guidelines—not legal advice—were offered in this chapter regarding when innovators should start to consider protecting their ideas. The innovator who implements a scientific method for innovation already enjoys a significant advantage over innovators caught in the pure art of innovation.

TAKE AWAY

1. An innovation is practically worthless unless it is adequately protected.
2. Protection of innovation actually fosters and promotes more innovation, so the paradigm shift from art to science

is a self-fulfilling prophesy, with greater innovation protections acting as a feedback loop to accelerate the pace of new innovations.

3. Once protected, there are various ways to exploit an intellectual property asset, principally licensing and litigation.

4. Four basic tools exist for protecting an innovation: patents, trademarks, copyrights, and trade secrets.

5. A patent owner does not actually have to make its patented product or use its patented process. This means that an innovator can develop an innovation and license someone else to make and sell it.

6. The term of a patent is fixed to no more than twenty years from the earliest effective filing date of the patent; hence the term 'limited monopoly' is used.

7. Certain innovations, such as music, photos, movies, software, and literature, enjoy automatic copyright protection if they are original works.

8. Legally, innovation includes two components: invention and reduction to practice. Invention is the formation in the mind of the idea with enough detail to build it. Reduction to practice is basically the construction or implementation of the idea.

9. Ideas are inherently intangible. Therefore a critical part of the patenting process is to document these ideas on paper. The unique patent system of the U.S. rewards the first to invent—not the first person who rushes to the Patent Office with a filing.

10. An inventor has an important obligation to disclose all prior art known to him or her but has no obligation to conduct a search for prior art. The Patent Office will do its own search, which is part of the application fee.

11. One strategy, particularly if cost is a consideration, is to file something called a provisional patent application. These applications can be converted within one year to a normal patent application that is then examined.

12. Changing the paradigm of innovation from pure art to science naturally leads innovators to think about intellectual property protection primarily as a self-preservation response.

13. Artistic innovators (art) enjoy automatic protection for their mental output, but technological innovators (science) must take affirmative steps to protect their mental output.

14. The framework for innovative thinking set forth herein will naturally cause innovators who follow this framework to think about protection, and thinking about protection is nine-tenths of the battle won.

COMMERCIALIZING THE INNOVATION

WAYNE ROTHSCHILD

Statistics indicate that only one in every twenty-five patents has a commercial value. As the process for patenting costs around $15,000 each time, an organization could probably save the cost of patenting the other twenty-four if it knew that they do not have the potential to achieve commercial success. The proposed CAISH model provides a structured framework to economically evaluate the commercial viability of innovations and manage the associated risk. When creative talent is energized, a number of innovative ideas can be generated very quickly. However, it can take months or even years to convert an idea into a commercially successful invention. While ideas are free, an invention is not.

INTRODUCTION

Patents are perceived in terms of technology and law. This thinking must change. Patents must be thought of as business first, law second, and technology third.

For example, the United States Patent and Trademark Office has issued more the 6.5 million utility patents since its inception in 1790. At the present time, the average cost to file a utility patent ranges from approximately $10,000 to $20,000. Assuming that all these patents were filed at today's average cost of $15,000, that is an investment of 6.5M x $15K = $97.5B in patent drafting and prosecution costs alone. This figure does not account for the cost of developing those patented products and technologies, which far exceeds the cost of the patents.

Only 4% of patents have any estimated commercial value. If this is true, only $3.9B of the dollars spent patenting these ideas can be considered investment grade.

This chapter focuses on the commercialization of patents and the patent licensing process. In other words, this chapter highlights concepts and a process for managing opportunity and financial risks when commercializing ideas. It teaches a step-by-step method to ensure that technical and legal resources are spent appropriately and driven by a financial model.

A part of the American dream is to invent something, patent it, and make millions of dollars. As a result of this "get rich quick" mentality, individuals and corporations are willing to invest capital and labor into patenting their ideas. In reality, transforming an idea from a patent into hard currency is challenging.

When an inventor has an idea, he/she typically runs to the garage, basement or R&D lab in an attempt to invent it (legally referred to as "reduction to practice"). The invention process can be all-consuming; some inventors quit their jobs; some corporations reassign the individual or team to the lab full-time.

Once the concept is reduced to practice, the inventor typically runs to a patent attorney. The patent attorney performs an initial search of patent prior art and begins preparing a patent. After the patent is filed, commercialization efforts typically begin, which means investing time and money into a dart-picked stock without understanding what business the company is engaged in or performing any research into profitability or growth projections.

THE CAISH MODEL FOR INTELLECTUAL PROPERTY PROFITABILITY

The CAISH Model for Intellectual Property Profitability was developed by Wayne H. Rothschild at Innovation Management, LLC in 2004 (www.im-llc.com). It is a method to evaluate the financial merit of an idea before the investment is made to develop the technology. It is also a method to evaluate the commercial viability and value of existing patents and portfolios.

For a new idea, the CAISH process is:

Conceive It → Assess It → Invent It → Secure It → Harvest It.

CAISH MODEL IDEA EVALUATION STEPS

1) *Conceive It—conceptualize the idea*
   Document the idea
2) *Assess It—do the math to make sure the idea has financial merit*
   Develop preliminary assumptions of market size
   Calculate the *ball-park* expected value of the idea
   Estimate the capital needed to develop and implement the technology
   Determine if necessary capital is potentially available
   Research patent and non-patent prior art
   Map the market, identifying the consumers, collaborators and competitors
   Calculate a budgetary valuation of the idea's profitability
3) *Invent It—develop the technology, reduce it to practice*
4) *Secure It—protect it*
   Protect the Technology
5) *Harvest It—make, use, sell or license it*
   Perform a detailed valuation
   Strategize the commercialization of the intellectual property (IP)

When working with an existing patent or portfolio, the process is:

Define It → Value It → Commercialize It.

CAISH MODEL PATENT AND PORTFOLIO EVALUATION STEPS

1) *Define It*
   Categorize the portfolio and broadly define each group of patents
2) *Assess It*
   Develop preliminary assumptions of market size and value for each group

Calculate the expected budgetary value of the profitability of the IP

Prioritize the opportunities using a *profitability index.*

Research patent and non-patent prior art

Map the market consumers, collaborators and competitors

3) *Harvest It*

Perform a detailed valuation

Strategize the commercialization of the IP

Proper management of an intellectual property portfolio is a major undertaking. Most large corporations have invested hundreds of millions of dollars in research and development and have spent millions more in the protection of their intellectual property. According to Sarbanes-Oxley, the corporation is to be held responsible for properly managing its assets. Accordingly, the CAISH process requires that management of this IP portfolio must not be left exclusively to the lawyers. A concerted effort must be undertaken to extract value from these IP assets, and this effort must be managed by a team consisting of business, technical, and legal professionals. Usually, the team needs a quarterback/project manager/facilitator to manage the project as well.

IP commercialization teams must assess the IP inventory and identify those assets which can generate revenue. Unmanaged portfolios are typically ripe with patents that are being violated and patents for which maintenance fees are being paid unnecessarily. In large portfolios, opportunities always exist where out-licensing can generate additional revenues. In most companies, the scientists turn the innovation over to the attorneys for patenting, and thereafter, an actively-engaged business perspective is rarely used to manage the IP assets. Using the CAISH model, a small focused staff can extract CAISH from assets the company already owns, resulting in a significant enhancement of incremental revenues from R&D with very high profitability.

After IBM began to focus on licensing as a business, it grew its licensing income from twenty million dollars in 1987 to over a billion dollars annually in 1997. This growth represents a consistent yearly return of more than one dollar per share. By

2000, the annual return grew to over $1.7 billion[1]. Developing a disciplined team to manage an IP portfolio can generate a tremendous return on investment.

PATENTS AND IDEAS

New ideas and/or inventions can come from anywhere. Sometimes a researcher invents something while trying to develop something entirely different. Sometimes researchers work their entire careers before completing an invention. Still other times, an individual invents something to solve a problem or invents something related directly to business needs. Regardless of the path to invention, the success of the invention is judged on its importance to the world, the ability to convert the invention into CAISH, or the ability to keep competitors out of the market.

IDEAS ARE CHEAP

Countless times people see a product and say, "I thought of that product years ago but never did anything about it! I could have sold that idea and made a million bucks." A famous and very creative video game designer, Al Thomas says, "I can develop an idea in my head in about ten seconds. But it takes the team a year or more to implement that idea. I should only come to work for about 10 seconds a year." The heavy lifting is in reducing the idea to practice. Although the inspiration is very important, the value creation is in the execution of the idea. Over the years, Al Thomas has been heavily involved in the creative implementation, execution, and commercialization of his ideas. In reality, he works full days like people everywhere do.

VALUE IN KNOW HOW

Although ideas are cheap, the intimate knowledge in a field or expertise with a given technology is also considered intellectual property. This IP can also be sold and licensed independently or

[1]Fairfield Resources International Licensing Program. [Cited 20 April, 2005]. Available from http://www.frlicense.com/license.html.

as part of a patent licensing agreement. Typically, an hourly or daily rate is paid for the expertise in a consulting arrangement. Other arrangements are beyond the scope of this chapter.

INVENTOR'S DEMENTIA

When inventors have an idea, they are typically very proud of the idea and would benefit from technology commercialization experience to humble and balance that pride. Friends and family are supportive of the inventor and the idea, and this support can create a false sense that it is a good idea. Business modeling and various methods of market research are necessary to ensure that the idea is in fact worthy of pursuit. Firms, such as Glenview, Illinois-based Gerald Linda and Associates (www.gla-mktg.com), are available to perform such verification market research.

FUTURE VALUE OF MONEY

Most companies use discounted CAISH flow analysis to determine the present value of future CAISH flows. This means that a dollar today is worth more than a dollar tomorrow. As such, the future revenues generated by commercialized intellectual property must be discounted in order to determine the real rate of return. Determining the net present value of the project is the responsibility of the financial member of the team. In the examples that follow, present value is ignored for the sake of simplicity.

COMMERCIALIZATION

According to Eliyahu M. Goldratt, *the goal is to make more money*[2]. Converting the invention into money can be accomplished through license, sale, donation or production. Licensing of the intellectual property is when fees are paid by another entity to the licensor to make or use the intellectual property, and the title is retained by the licensor. Sale occurs

[2] Goldratt, Eliyahu M. and Cox, Jeff. The Goal. Great Barrington, MA: The North River Press, 1992.

when the title of the intellectual property transfers to the buyer. Both sale and license agreements can have royalty components. Donation is when the intellectual property is donated to a university or other charitable organization, and the donating company thereby reduces its tax burden. Production is when the company finds a way to manufacture or use the invention for competitive advantage.

Corporate strategy and business modeling are the keys to deciding what method or combination of methods to use in the commercialization of the IP. If the technology is core to the corporate strategy, many companies choose to keep the IP for internal use only. Other companies will license the technology, but only for use outside their industry. Still other examples abound in the high technology sectors where cross licensing is the key to component standardization, which intensifies competition and pushes production costs down.

Non-core IP technology can also be used by the company, or the company can spin-off a new company or a new division to focus on the exploitation of the IP. Be advised that most researchers and innovators may not make the best business managers. As a result, the leadership of the new entity may not be the person that invented the technology.

Non-core technology is usually ripe for license, sale or donation. If the technology does not have sufficient value, the patent maintenance fees should not be paid. Well-managed IP portfolios are reviewed on a regular basis, and the patents are categorized and licensed, sold, donated or deleted accordingly.

The CAISH method that follows is an important step in evaluating the IP and determining the strategy for the particular asset(s). This method applies both to new ideas and existing portfolios.

CAISH AND THE ECONOMICS OF COMMERCIALIZATION

As discussed previously, inventors and corporations have a dismal success rate in the conversion of intellectual property into

CAISH. A very well-planned process is needed to evaluate ideas prior to invention.

While ideas are essentially free, invention is not. Invention requires a substantial investment of time, money and other resources. CAISH is a logical approach where business modeling is required. In fact, the first step is to understand the balance of risk and return, especially with respect to the opportunity cost of investing in something else. This understanding requires a thorough analysis of the likely best case, worst case, and middle case outcomes regarding the time and money to be invested and the potential returns. This analysis is based on the principle that business modeling is quicker and much less expensive than the research and development required to invent.

In the traditional innovation model (see Figure 13.1), investment leads the generation of value. In this case, the time, money and resources are spent inventing and prototyping before the market value is even estimated. The CAISH model (see Figure 13.2) is designed to invest in the project *while value is being created*. The CAISH method is a structured approach which maximizes return while mitigating risk. CAISH may increase the rate of return on R&D and helps to increase the odds that the creative researchers are working on the most lucrative projects.

As Figures 13.1 and 13.2 show, the idea has very little value until it is a unique, protectable and commercializable technology or product. In the CAISH model, the transformation of an idea into an invention into a product is shown by climbing the *expense mountain* (the Cost Curve). The *value creation curve* (Value Curve) shows that value is created through the development, marketing and selling of the product.

Intellectual property can be sold anywhere along the value creation curve, but generally the value to be extracted from the sale increases along the curve. However, the risk adjusted return may be higher if the transaction occurs on a lower point in the mountain, especially in litigation where the defendant may have infringed on the patent owner's claims. However, the total dollar value of the transaction will be lower, especially when IP is sold to an entity that is better or faster at climbing the value creation curve. For example, when a start-up

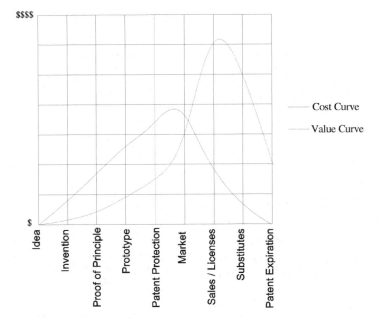

FIGURE 13.1: The Traditional Innovation Model

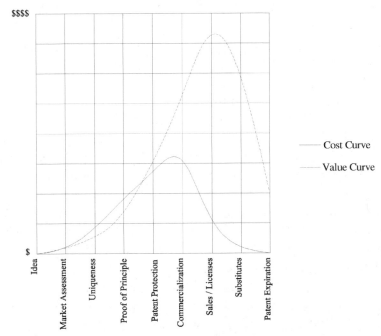

FIGURE 13.2: The CAISH Model for Intellectual Property Profitability

attempts to develop a new pharmaceutical compound, the staff and laboratory requirements may be excessive, while a large pharmaceutical company may be able to perform the initial evaluation with virtually no incremental direct overhead. However, the opportunity cost for a large pharmaceutical company may be considerable.

EXAMPLE

To better understand the concepts in this chapter, the following highly simplified and fabricated example will be used.

Assume that Eraser Head Corporation owns the foundational patent for the elastomer (rubber or other similar materials) eraser which is integrated in the end of any wooden or mechanical pencil. This patent will be referred to as the *Integrated Eraser Patent*. Also assume that the patent for the wooden pencil has expired, but the patent for the mechanical pencil is still in force and owned by Clicker, Inc. Woody Ltd. is one of many manufacturers of commodity wooden pencils.

CAISH EVALUATION

Even before a person or corporation begins developing an idea, an assessment of the idea and the ensuing business opportunity must be ascertained. The CAISH steps are a framework for the assessment of the idea and its commercialization. A serious inventor or business analyst will use these steps to create a document which defines the idea. The document will be the road map and the information source for the commercialization effort. It will outline the idea, features, value, markets, competitive position, protection, and commercialization strategy. Of course in actual implementation, the document is a living document designed to take into account the changes in the commercialization environment going forward.

While this thought process will be highlighted using the Integrated Eraser Patent example, the process applies to a technological breakthrough, a business method, a medical advancement and other intellectual property. Addressing all of the steps is a time-consuming exercise. From a risk-reward perspective, doing so will pay huge dividends, as the time and cost to perform the analysis is miniscule compared to the time and cost of actually developing the technology and protecting the idea.

CONCEIVE IT

Document the Idea

Defining what the idea is and what it will do is an important first step. This step is necessary for new ideas, existing patents and portfolios. Because this step forces the inventor (or the business manager) to focus on the idea and its applications, many ideas are dismissed in this initial process. The definition can be a paragraph or a few pages. Generally the definition is one to two pages, and many parts of this document will be subsequently used as a basis for disclosure documents, patent filings and business plans.

Some of the best inventions solve a problem which companies or individuals are currently experiencing. Be certain that the definition of the problem and the solutions are included in the document.

If the idea is a new one, the document should be signed and witnessed. Depending on the nature of the industry, filing the idea with the United States Patent and Trademark Office under the disclosure document program http://www.uspto.gov/web/offices/pac/disdo.html, where it can be used to prove the date of first conception, may be prudent. The disclosure document should teach someone of ordinary skill, who is skilled in the art, to build the invention.

Ownable Distinction

A specific product feature can be used to differentiate one product from another. When the importance of a specific product feature is high enough, purchase decisions are driven by the presence or absence of this feature.

If the feature can be protected through patents or trademarks, then that specific feature can provide *ownable distinction,*[3] where the feature is unique to a particular brand or a particular manufacturer. Using trademarks protects the name and image (brand) of the feature, while using patents protects the utility, design or method of using the feature. Because purchase decisions are based on this feature, the value to be gained from *ownable distinction* can be immense.

Thinking back to the basics of patent law, owning the IP for a highly desirable feature on an existing product or technology can be sufficient to unlock real incremental value, even if the foundational technology is owned by someone else.

ASSESS IT

Preliminary Assumptions

The first pass assumptions are *ball park* estimates and are typically based on limited research. Most of the information will come off the top of peoples' heads, and the data points that stand out should receive a sanity check. Preliminary assumptions provide a quick first evaluation of the idea. Although little research is performed, at least three experienced people, representing the technical side, business side, and operations side, should review the assumptions before moving forward. The data should be entered into a financial projection spreadsheet, which can be easily changed as the data become more accurate. The preliminary assumptions are a guide regarding what data must be researched and what the value of the IP may be.

Accountability and Risk of Failure

Since the technology has not yet been developed at this stage, risk of failure is still a reality, in which case the entire invest-

---

[3]Cahr, Darren S. and Morgan. Mary T. "Got Ownable Distinctiveness?" Brand Packaging, September/October 2001, 10.

ment may be lost. Assumptions are never accurate, but they must be reasonable.

The key to managing these estimating techniques is accountability. The team that created the estimates must be held accountable for delivering reasonably accurate rates of return for the project. For this reason, these forecasting tools must be used by a team consisting of members with sales, marketing, operations, finance, legal, and technical experience. If in reality the numbers are terribly inaccurate, then the variances must be analyzed and the learning applied to existing and future analysis on other projects. Because big estimates can be used to sway a project, the ability for individuals to game the system and be mitigated by accountability is critical.

Expected Value

To develop market assessments and determine risk adjusted returns for a given piece of intellectual property, assumptions must be created. The more experience with the market one has, the better these assumptions will be. In any case, a series of best and worst assumptions must be created, and then likelihood should be assigned to each assumption. The expected value (weighted average) of the forecast can be created by adding together the product of each assumption and its likelihood. The Integrated Eraser Patent is used in the following example.

Estimated Total Market Wood Pencils   100 million units
Potential Market Share for Eraser Head Pencils

	Share	Likelihood
High estimate	55%	15%
Medium estimate	35%	45%
Low estimate	15%	40%

Expected Unit Sales
  $= 100$ million $* (55\% * 15\% + 35\% * 45\%$
           $+ 15\% * 40\%)$
$E(x)$units $= 30$ M units

Potential Incremental Gross Profit for Eraser Heads
(Incremental price—incremental cost)

	Incr Profit	Likelihood
High estimate	$ 0.05	30%
Medium estimate	$ 0.03	50%
Low estimate	$ 0.005	20%

Expected Incremental Price
= $0.05 * 30% + $0.03 * 50% + $0.005 * 20%
E(x)price = $0.03

The *ball-park* expected value for the gross profit is:
30 M units * $0.03 = $ 900 K

Development Costs
Potential Costs to Develop the Technology

	Dev Cost	Likelihood
High estimate	$ 1 M	20%
Medium estimate	$ 500 K	50%
Low estimate	$ 250 K	30%

Expected Development Cost
= ($ 1 M * 20% + $ 500 K * 50% + $250K * 30%)
E(x)development cost = $ 525 K

The use of expected value balances the optimistic and pessimistic views of the project. The range of outcomes far exceeds the expected value. For example, based on unweighted assumptions,

Maximum Gross Profit: 100 M * 55% * $ 0.05 = $ 2.75 M
Minimum Gross Profit: 100 M * 15% * $0.005 = $ 75K

Comparing the expected value of the cost with the expected value of the profit, the project looks profitable. However, at the extreme case of cost and profit, this project is a loser. Depending on the opportunity costs, this project may or may not be worth the green light. Taking the present value of the CAISH flows can also reduce the value of the opportunity.

In some cases, the patents may be filed in order to provide the entity with the option to pursue or preclude others from commercializing the product or technology at a later date. This is generally referred to as Option Value.

In the product planning cycle, multiple projects may be analyzed in this way prior to a project review meeting. The project risks and returns can be compared, and the best projects can be green lit. This approach is discussed in more detail later in this chapter.

Of course, all of the above numbers are based on assumptions which have not been verified. The next step is to begin to challenge the assumptions to see if the volume can be increased, the price can be increased, or the developmental costs can be reduced. The amount of time spent verifying the assumptions will depend on the perceived validity of the numbers. In the base case, the numbers can be compared with the historical performance of similar product enhancements. At the other extreme, actual market research can be conducted to make such a determination. The investigation procedure is entirely based on the risks and returns involved in the further development of the project.

Research Prior Art

To perform a background investigation, use the Internet, the telephone and hands-on evaluation of relevant products and technology. Use market research to search for similar products, technologies and alternate solutions to the same and similar problems. Review research journals, trade journals and consumer publications in the search for prior art. Keep good records regarding what is learned, the industry players and the end users.

Search the United States Patent and Trademark Office (www.uspto.gov) for patent prior art and related patents. Although the US patent website offers free access to the database, other subscription services such as Delphion (www.delphion.com) offer much more comprehensive and faster research tools. The efficiency to be gained through the use of these tools can more than justify the subscription costs, which are available on a monthly or annual basis. Subscription-based prior art searches and defensive publications are also available at www.ip.com.

Keep an organized record of the journals and databases searched, patents reviewed and documents identified. Draw a map of the entire space. Even if similar technologies are found, record the strengths and limitations of each. Remember that *ownable distinction* can be achieved with the addition and protection of one key feature or important point of differentiation. Additionally, the process of identifying what exists is often what spurs the breakthrough idea. If this research is done before the invention is complete, it will guide the invention process to create something that is unique and protectable.

During the research, keep lists of product ideas and licensing opportunities. As each competitive and collaborative technology is identified, be alert to the companies and industries that are developing and using the technologies. This will identify potential licensing-in and licensing-out opportunities.

When evaluating the patents, be acutely alert for foundational patents. These are patents which have numerous forward citations and are typically the breakthrough patents which define the technology. If the claims are strong, these patents typically constitute very high value intellectual property. If the claims are weak, these patents open the technology, and the more recent patents typically protect the specific applications of the technology. Keep this in mind during the protection phase and strive to create foundational patents.

If a patent (or patents) is (are) found, which closely resemble the technology and intended applications but do not appear to be actively utilized, this situation may be an indication that little value is present. Although a cautious approach is required, consider contacting the inventor (checking with the attorney first) to determine the status of the intellectual property. Note that the inventor's and the patent attorney's contact information is a matter of public record and is available at www.uspto.com.

Most people are willing to share their experiences. Some inventors are waiting for licensees to find them, while others are out beating the bushes. Use caution with this approach, as you do not want to be found willfully infringing on someone's patent, nor do you want to disclose your idea to them and risk that they file a continuation of their patent which includes your idea. Recognize that in these cases, licensing opportunities abound. Combining networking with Internet research is far more powerful than the Internet alone.

Map the Market

Studying and understanding the markets, the substitutes and the competitors is critically important. Substitutes explain how people manage to address this problem today, without the invention under consideration. Consider the cost of the existing solution as well as the time and effort to use the current solution. Does the innovation rectify a problem with the existing solution? Is the difference meaningful? Does the new solution fit within the user's current and existing behavior? Can the current solution incorporate the innovation under consideration? Changing behaviors can be a more significant obstacle than one would expect.

Competitive products, and the competitor's response to the invention, must be considered as part of the business plan. Is the industry keen on patents and intellectual property, or is IP new to this industry? Does the competitor possess prior art? When competitor 1 moves to enforce a patent on competitor 2, might competitor 2 have patents which they will be encouraged to enforce? The ramifications of any IP action must be considered before taking action. Licensing, cross-licensing, litigation, and obstructing competitors are all possibilities.

Profitability Index

Using the market analysis, prior art research, and financial forecasts, a reasonable financial model can be created for the project. Rate of return, return on investment, and expected dollar value of profit, either for the entire project or for milestone options, can be used to compare the various opportunities. If a particular bottleneck resource exists at the company, then create a *profitability index* for each project. The profitability index for a given project is determined by dividing the expected profit from the project by the cost of the bottleneck resource. All projects can be compared, and the projects that provide the best return on the bottleneck resource can be identified. In the simplest example, assume a software company has a bottleneck in software engineering. Assigning the software engineers to the most lucrative projects will maximize overall profitability.

The following example explains how to determine the cost of the bottleneck resource. Assume that the bottleneck is engineering. Select a dollar value for one hour of engineering and multiply it by the number of hours needed for a given project. As long as all projects use the same engineering hourly rate, then the rate is irrelevant. Divide the expected profit dollars from the project by the cost of the bottleneck resource to determine the profitability index. The profitability index of all projects can then be compared.

Note that the profitability index method ignores strategic reasons related to specific projects, and it also ignores the magnitude of the profitability. The data must be reviewed with these limitations in mind, and decisions must be made accordingly. With these factors taken into account, this method is an excellent way to review the opportunities and maximize profitability.

SECURE IT

If the financials justify it, then invent the product or the technology. Use milestones to track and manage the progress according to budgets and schedules. Creativity can be scheduled, and the search for revolutionary discoveries can have deadlines. These deadlines and schedules are driven by the financial model.

The project was green lit based on a schedule and a cost. At any point, if the projected costs and/or time frames are expected to reach beyond the model, then the analysis must be reviewed, and the assumptions must be challenged. With the new analysis in hand, the project may need to be cancelled, which is consistent with both an Options-based and an end-to-end project model.

Aside from medical research developing an important breakthrough, and individuals inventing to solve a personal need, innovation is generally driven by companies searching for new ways to increase profits. As such, it must be justified by the financials and monitored for compliance to the plan.

Protect the Technology

When writing a patent to protect an invention, understanding the way it will be used is important. The best first question is, "Upon whom will you exercise the intellectual property?" This question is meaningful because a patent must be enforceable. Exercising a patent against a customer, the government, a non-domestic competitor or a charity may be difficult legally and politically, and it may be very costly. As such, very little value may exist in patenting such an invention. Defining how the patent will be enforced will also affect the way the claims are crafted.

International patents are valuable but very expensive. Translation costs, filing fees and maintenance fees in each country add up very quickly. As a result, a global patent strategy is important. Even if someone in Slovakia may knock off the invention, are they likely to cause sufficient harm? Is the risk-adjusted cost of the patent worth the protection it provides?

For example, if the product is expected to generate $50K in revenues and $15K in profits in Slovakia over the next few years, it is not worth the cost of the patent (which may be $10K) in that country. Even without patent protection, the technology can still be sold or licensed into that country, and potentially generate $30K in revenue and $8K in profits even with the competition (but without the patent expense).

Generally, protecting the technology in countries where significant opportunities exist and where potential licensees (competitors) have significant operations is the rule of thumb. When the innovations are in a capital intensive environment, patents are only needed in those counties where competitive products are manufactured, assembled, and widely sold. For the automotive industry, this includes approximately ten countries. In this case, competitors outside of those countries do not have the scale or resources to be a competitive threat. Therefore the risk is not high enough to justify the cost of those additional patents.

Basics of Patent Law

As discussed in chapter 12, patents are opinions that the invention is unique and as such are still subject to being challenged. Patents do not give the explicit right to *make, use or sell* the invention; they only prevent others from being able to *make, use, or sell* the invention within the national patent's protection, until the patent expires or until it is found to be invalid.

An issued patent is not guaranteed protection of an idea. The patent is only an opinion from the patent office that the invention is novel. If relevant patent or non-patent prior art can be found, the courts may find that the patent is not valid. This element of risk must be taken seriously, as it can undermine all intellectual property valuations.

Oftentimes, a previously issued patent is the basis for a new invention (the new patent is seen as an improvement). In this case the new patent holder cannot practice his or her invention without a license to the prior patent, or until the prior patent expires. In addition, the original patent holder cannot practice the specific newly patented improvement without a license from the new patent holder.

If an inventor gets specific enough, almost anything can be patented. However, a patent that is very specific (very *narrow*) can typically be circumvented (a process called *design around*) by avoiding the specific patented feature.

Strategic Value of Intellectual Property

Intellectual property is not always about making the product. Often an innovation will occur which is not along the direction of the company's technology, but it may lie in the road ahead for the competition. When this occurs, the company may be well advised to patent this technology as a strategic tool to use against its competitor.

For example, a wire line technology company happens into technology that is likely to be important to wireless companies a few years from now. Even though the wire line company does not plan to fully develop or utilize this technology, it may invest in the patent, because it may be able to license or cross-license this technology at a later date.

Strategic-value patents are high risk, high return opportunities. If the perceived breakthrough will be fundamental to an industry, then by all means, patent it. If an assessment of the technology reveals that multiple ways exist to solve the same problem, then either patent all of them or take a pass.

Option value patents are also long term bets. Strategically, an entity may choose to file a patent on a technology for the purpose of having the option to commercialize the intellectual property at a later date, or to prevent others from commercializing the intellectual property. The cost of the option is the cost of developing and patenting the technology. The value of the option will be determined over time.

HARVEST IT

Detailed Valuation

The value of anything is the price at which the seller is willing to sell and the buyer is willing to buy. Typically, valuation of intellectual property is calculated using a blend of the three accepted methods: cost, income and market. None of these methods should be used as a stand alone. Logical (or illogical) factors are always present which affect the buyer's willingness to pay and the seller's willingness to sell. Keep the emotions off the table. For example, a given inventor has CAISH, time, blood, sweat and tears invested in the invention. While this investment does not affect the buyer's valuation of the invention, it does affect the inventor's perceptions and expectations. On the other side of the equation, if the innovation has strategic value for the buyer, then a premium should be considered.

Cost Method

The concept of cost includes the estimated cost to develop and protect the technology and the expected cost to design around the patent. Note that in reality, the development costs are sunk costs, and therefore any MBA will assert that they are not consequential to the negotiation. However, in reality, everyone wants to see a return on their investment, and the development cost is seen as an investment.

The cost to design around the patent is a make/buy decision. It must include research costs, time to market considerations, risk factors for potential failure, and risk factors for potential patent litigation. Furthermore, the opportunity cost of spending resources to develop technology which can be purchased is always present.

Referring to the pencil eraser example, assume that two engineers spent six months and $100,000 developing, prototyping and testing the technology. Eraser Head also paid $20,000 to patent it. Therefore, assume that the cost to develop the technology was approximately $250,000.

Woody, the pencil manufacturing competitor, is considering a license. Woody believes that its engineers may be able to develop some type of competing technology. Their ideas include 1) pencil markings that rub off with your hand, 2) pencil markings that are erasable with wood from the pencil and 3) special paint on the pencil that will erase lead. Woody estimates that any of these ideas will require 2 years to develop, have a 20% likelihood of success, and cost at least $500,000 each. In addition, very little risk of infringement exists with any of these three ideas. As a result, the cost to design around the patent will range from $0.5 to $1.5 million, and the success rate of developing the technology will only be 20%. With a 20% likelihood of success, this research would need to be conducted five times to have a *sure-thing* solution. As a result, the risk adjusted design around cost ranges from $2.5 million to $7.5 million.

In this example, the cost method value has a very wide range:

Development Cost	$250K
Design Around Cost	$ 0.5 M to 1.5 M
Risk Adjusted Design Around Cost	$2.5M to 7.5 M

Income Method

Determining the present value of the economic benefit afforded by the patent is a second method of valuation. The objective of the income method is to determine the incremental income that is created by way of the technology.

In the case of the pencil eraser example, assume that eraser-less pencils cost one cent to make and sell for two cents. Pencils with erasers will sell for five cents. The incremental costs, including manufacturing, materials, labor and amortization of the manufacturing equipment, adds one cent to the cost of the product. Therefore, two cents of additional margin are in every new pencil sold.

Assume that the market consumes 100 million pencils without erasers annually and that given the choice, 30% of these consumers will buy the pencils with erasers at the above prices. The incremental income due to the eraser pencil will be 100 million pencils * 30% * 2 cents margin = $600,000. A five-year license to the patent would therefore generate $3 million of incremental income for the licensee.

Pencils with erasers may attract new consumers and increase consumption by existing users. To this end, assume that the market grows by an additional 15% as a result of the integrated eraser feature. This creates the following incremental value: 100 million pencils * 15% * (1 cent pencil margin + 2 cents eraser margin) = $450,000. The five year license would therefore generate $2.25 million in incremental income based on the market growth.

Over five years, the integrated eraser IP is projected to create incremental income of $5.25 million. Note that this model does not account for any year-over-year changes in market consumption, ignores the present value of money and is limited to a five year window. Depending on the accuracy of the projection, these factors should be taken into account.

However, the patent holder alone may not have access to the marketing channels to sell the product. The company that deploys its working capital, marketing skills and sales force in order to capture 30% of the market is entitled to a significant portion of this income. Deciding on the relative share of these earnings is part of the negotiation.

The *Rule of 25* must be mentioned. Some licensing practitioners believe that 25% of the income generated from the patent is due to the inventor. The validity of this rule is often questioned. Royalties are whatever can be negotiated. In the toy industry, the basic rate is 10% of net-wholesale, unless the product will use the patent and a brand of toys. If that is the case, then the royalty may only be 5% of net wholesale, because the brand licensor must also be compensated. Getting fair compensation requires a strict definition of net-wholesale or whatever the payment basis may be. Remember that everything is negotiable. Make sure that the maximum value is extracted from every term of the deal.

Using the *rule of 25*, the licensor would be entitled to:

Annual royalty
($600,000 + $450,000) * 25% = $ 262,500

Over Five Years
$262,500 * 5 = $1,312,500

Remember that 25% may be too high or too low, depending on the value of the IP, the level of risk, and the costs to pioneer and commercialize the technology. Another major factor is the terms of the deal. License fees, advances on royalties, minimum guarantees and other factors will significantly affect the amounts to be paid. Each party to the negotiation has different needs and different abilities to pay. For example, a start-up company may be willing to pay a larger per-unit royalty, while a large corporation may want to pay a lump sum upfront with no royalty. Again, this determination is all subject to the negotiation. Generally, understanding each others' needs will help everyone get what they need at the lowest total cost.

Side Note:

The licensor should never base compensation on net income or net profits, because the profit number is easily manipulated by overhead allocations and other factors which will conspire to ensure that the royalty is zero.

The licensee should always try to base the license agreement and the royalties on net income.

Market Method

Benchmarking the licensing costs for similar technology agreements is another method used in the valuation of the IP. On the positive side, various surveys, publications and other sources of information exist about the terms of thousands of deals. On the negative side, no two deals are ever the same. Sources include, but are not limited to: royaltysource.com,recap.com and royaltystat.com.

Review the list of typical factors to get an idea of some of the variables. The published deal outcomes typically do not list all the terms of the deal. Attempting to value all the points of differentiation and compare them is no small feat. Furthermore, consider that many deals are done under duress from either the buyer or the seller. This information is not evident in the data.

Using the eraser example, assume that research found data on the following patent licenses. Note that the market assumptions and expected value projections are used to apply each of these deals to the Integrated Eraser Patent example.

1) A pen cap patent was licensed non-exclusively for $10,000 down and 1 cent per unit.

   5 years $*$ (100M units $*$ 30% + 100M $*$ 15% incremental sales) $*$ $0.01 + $10K = Equivalent value = $ 2.26 M

2) An eraser keychain patent was licensed exclusively for a $20,000 one time fee.

3) A pen clip patent was licensed non-exclusively for $50,000 down and 2 cents per unit.

   5 years $*$ (100M units $*$ 30% + 100M $*$ 15% incremental sales) $*$ $0.02 + $50K = Equivalent value = $ 4.55 M

These licenses are not the equivalent technology or application; however, they are industry examples with information on the prices paid.

DETERMINATION OF VALUE

With knowledge of the three valuation methods, a framework for the value can be created. Remember that understanding the aspects of the license that are important to each party is crucial.

Term sheets are documents which summarize the negotiating points of a sales or license agreement. Before beginning the negotiation, each party should create their private term sheet and decide who will write the working term sheet. The private term sheet defines the objectives to be gained from the negotiation. The working term sheet is the working summary of the negotiated terms. Using a term sheet will simplify the negotiation. And when the agreement is written by the attorneys, it will make the terms clear and significantly reduce legal bills.

Generally, small companies and inventors should let the big company open the price negotiation. What the big company perceives to be small dollars may exceed the small company's or inventor's wildest expectations. In any event, once a price is stated by the buyer, the seller will try to increase the price. Once a price is set by the seller, the buyer will try to decrease the price.

The contrary is also true; the big company wants to get the smaller company to open the price negotiation. In any case, someone will need to open the bidding or make the first proposal.

In the integrated eraser example, what should the price be? It depends on where the power exists in the relationship. Assume Eraser Head does not want to manufacture pencils but still wants to convert the Integrated Eraser Patent into CAISH. To do so, Eraser Head needs to license or sell the patent.

Woody wants to increase margins, volume and market share. At this time it has no patents to serve as barriers to competition. Woody desperately needs a way to differentiate its product from the competition, and this patent would provide *ownable distinction*. Woody wants an exclusive license to prevent the competitions from getting access to the Integrated Eraser Patent. Eraser Head has completed its due diligence and understands this value opportunity. Eraser Head can play Woody against its competition during negotiations of an exclusive license or for industry wide non-exclusive licenses.

In this negotiation, Eraser Head has great leverage. On the contrary, the wooden pencil trade association can join together to negotiate a license, in which case the industry group may have leverage. However, at least one company is usually willing to cut a deal outside of the trade group, and this fact somewhat mitigates the trade association's leverage.

Clicker has its own patent which protects its market. However, it is always interested in increasing sales and margins.

If Eraser Head wants to exploit its patent in the mechanical pencil market, it must license Clicker, which has power in the negotiation and will likely get a better deal than Woody. Note that Clicker will get the equivalent of an exclusive license on the eraser patent, because it controls the mechanical pencil market.

VALUE CALCULATION

As is typical, in our example, the valuation methods have a very wide range. Note that the risk adjusted design around cost is just for reference and will not be used in any discussions or negotiation.

COST METHOD
    Development Cost $ 250 K
    Design around Cost $ 0.5 M to 1.5 M
    Risk Adjusted Design around Cost $ 2.5 M to 7.5 M

INCOME METHOD
    Incremental Income (5 years) $ 5.25 M
    Rule of 25 (5 years) $ 1.3 M

MARKET METHOD
    Pen Cap License $ 2.26 M
    Key Chain $ 20 K
    Pen Clip $ 4.55 M

Many companies lack the expertise and do not perform good due diligence. As a result, a well-prepared negotiator will have the advantage and will also have basis for justifying its bids. The unprepared party will have difficulty substantiating its counter offers.

If Woody opens the price negotiation, it should start at $20 K with no royalty and base this price on the eraser key chain patent license. If Clicker opens the price negotiation, it could start at $10 K based on its exclusive mechanical pencil patent and the eraser key chain license. If Eraser Head opens the price negotiation, it should start at $5.25 M based on the incremental income on the table and argue that the sales and growth projections are low.

Obviously these negotiations are always interesting. The final outcome will be based on the need, purchasing power, and negotiating skill of each party. One further note is to research the opposition and their negotiator. Learn about their track records and negotiating techniques. Ask around; how much you can learn is sometimes surprising.

As stated previously, to get the best out of any negotiation, understanding one's goals and objectives going into the negotiation is critical. Write the term sheets first. Further discussion of negotiation techniques is beyond the scope of this chapter.

BUNDLING

Bundling of related patents can increase the value of the portfolio, but make sure that the maximum value is extracted. As discussed previously, different elements are worth different amounts to different licensees. From a contracting cost standpoint, however, bundles can be much more efficient to sell and manage.

Putting a value on a bundle of IP can be complicated. One method is to sum the value of the most important patents and adjust this with a premium. The premium attempts to capture value from the other patents included in the deal. Often, these other patents are picket patents covering specific applications. These picket patents may have no commercial value without the more foundational patents. Adding the individual valuations of all of the patents typically results in an outrageous and unreasonable amount. Additionally, the substantial time required to perform a detailed valuation of all the patents may not generate a good return on the time invested.

SELL/LICENSE BACK

Recently, a business model is being exploited called *sell license back*. For example, a company, Inventor Inc., sells a major patent or a bundle of patents to a firm, Magnet Inc., in an arm's length deal. As part of the arrangement, Magnet licenses the patents back to the inventor for a royalty. Magnet is also free to license the patents to others. Sometimes, the inventor's license is exclusive in the industry; other times it is not.

A sell license back transaction has many effects:

1) Creates reportable income for Inventor Inc. as a result of the sale of the IP.

2) Converts the patents, which are an intangible asset, into CAISH, which is a tangible asset on Inventor Inc.'s balance sheet.

3) Allows deductions of royalty payments as operating expenses on Inventor Inc.'s income statement.

4) Generates royalty income for Magnet, Inc.

In essence, depending on how the deal is structured, the transaction can be either a sale or a financing transaction for Magnet and the inventor. Be aware that these transactions can be very complicated regarding legal contracts, intellectual property, valuation, tax, and financial and execution details. While the commercialization opportunities allow for larger market penetration and thus greater returns, the scale required to support the cost and complexity is on the higher side of commercial licensing transactions. These deals typically start with IP in the 10's of millions of dollars. Two Chicago-based firms which specialize in such transactions are Creative IP Solutions (www.creativeip.com) and ICMB Ocean Tomo (www.oceantomo.com).

STRATEGY FOR COMMERCIALIZATION

Clearly, many companies develop intellectual property to differentiate their products, create barriers to competition and sustain revenue growth. The financial analysis is just as important to prioritizing those opportunities as it is for licensing programs. The commercialization strategy defined in this chapter is a marketing strategy and focuses on the licensing opportunities, not the internal make and sell option.

Negotiating a deal is much more complicated than establishing the price for the intellectual property. Many terms need to be negotiated, and each term will have a different value to each party. For example, a given licensee may want to have exclusive access to the technology. This exclusivity may prohibit or allow the original inventor to exploit the technology, which may include or exclude the inventor's further research and development in the field, subject to the terms of the deal. On the other hand, the agreement may give the inventor the right to continue his or her research, but the licensee may automatically have ownership or the first right of refusal to negotiate a license to this new technology. Furthermore, the exclusivity may be for a period of time or for a geographic location or for a particular industry. The permutations of licensing options are endless.

Sometimes, a licensor and licensee may be unable to reach an agreement on price, because the licensee does not need the entire patent; the licensee may want exclusive access to only one or two claims. The license can be sliced up to fulfill this need by licensing only the claims that particular licensee requires, and licensing the remaining claims to others.

Understanding the licensee's inner workings is important here. At most companies, the legal department works out the details of the licensing deal. R&D and product development are typically only involved peripherally. As a result, the engineers may believe that they have license to the patent, when in reality they only have license to two claims. Additionally, employee turnover in both the technical and legal areas helps insure poor communication and a lack of knowledge sharing. As a result, future generations of product may violate the unlicensed patent claims. In these cases, there is a relationship and some precedent for the value of the patent claims; however, the licensor must be alert and identify the infringement.

Furthermore, at many companies the patents are in a file cabinet, and the researchers that invent the technology are not in touch with the market. As a result, many patents are not enforced because not enough value is perceived by the core business to divert resources to enforcement management.

In addition, many larger firms are more interested in having the Freedom to Operate in the space mapped by the patents. In this situation, patents are viewed as defensive and can be used to counter sue. An environment of litigation risk avoidance, which effectively prevents enforcement activities, develops among the major market share owners. Audits related to licensed rights are a recommended provision in an intellectual property agreement.

The IP will have different values to different companies and different industries. Take the time to map out the space and determine which company in which industry to approach first. Timing is everything. Companies tend to have a strategic focus as they try to grow a given market or enter a new market. Approaching a company with the right idea at the wrong time will not be successful.

For example, if an incumbent shoe polish company with good products and great market share exists, it may be less interested in a new breakthrough shoe polish technology than another big company which is trying to break into the shoe polish space. Do not be fooled by the large share; the product may be more valuable to the small player. The small player may be willing to pay a higher license fee with lower volume; thus more profit dollars may be gained in such a scenario. Once again, research is the key. Call and talk with the potential licensees. It is amazing how much can be learned and how much information people are willing to share. Even non-confidential information can be very enlightening.

When negotiating a licensing deal, sell each component to someone who values that specific component. Never give anything to anyone who is not willing to pay for it. For example, if the technology is broad, and there is an automotive company that is interested in an exclusive, negotiate an exclusive in the automotive industry. Then sell another exclusive in the aerospace industry. In this case, the automotive and aerospace licensees are non-competitive and may not be concerned about each other. By determining what is important to each party, potential license revenues can be doubled or tripled.

In another example, a food manufacturer may be willing to pay a unit royalty of ten cents per unit for exclusive access to an additive, of which they plan to use one million units/year. On a non-exclusive basis, this manufacturer is willing to pay five cents per unit. If additional licensees are interested, are willing to pay five cents per unit non-exclusively, and if the total volume of the additional licensees exceeds two million units/year, greater revenues will be generated by offering the technology non-exclusively. Note that non-exclusive licensing makes the licensor less dependent on one licensee both in terms of market penetration and ability to pay. The fate of large or small companies in today's swiftly moving markets is often changeable.

Other factors for licenses include end provisions whereby there are minimum performance guarantees to maintain exclusivity. In other situations, the rights revert to the licensor under certain non-performance conditions. Furthermore, in some cases, the licensee has the right to *pay-up* to satisfy the minimum guarantee and maintain the exclusivity.

Situations exist where a licensee is acting to prevent a product from coming to market. If the inventor has an interest in seeing the product in the market (i.e., the inventor sells an adjunct product), then *pay-up* terms are not a good idea. If the inventor is only interested in the CAISH, then the *pay-up* option is not problematic.

Typical factors to be negotiated in patent, trademark, trade secret and know-how licensing agreements include, but are not limited to:

> Exclusive rights (or exclusivity limitations)
> Geographical rights
> Industry applications
> Duration of license (term)
> Specific applications or uses
> Upfront payments
> Annual payments
> Royalty amounts
> Minimum guarantees
> Fees for technical expertise or support
> Rights to continuations, continuations in part, and/or
>   divisional applications
> Right to enforce and proceeds from enforcement

Right to sub-license or cross license
Restrictions on design-around
Restrictions on challenges to patent validity
Terms upon invalidation of the patent
Terms upon expiration of the patent
Confidentiality
Reporting rights
Payment terms
Auditing rights
Non-performance terms
Terms upon cancellation of the license
Methods of recourse

The more complicated a license becomes, the higher the contracting costs are likely to be. These costs are the costs and time required for the lawyers, and business and technical people, to hammer out and close the deal. Additionally, compliance with the terms of the deal must be able to be audited to make sure that all parties get a fair deal.

STRATEGIC PRICING FOR SUCCESS

When selling a license, many factors must be considered. The prospective licensee will consider the need for the license, the validity of the patents and the cost to design around the patent. Often times, licensors will price the patent such that a license is the least expensive option. As a result, the licensee is encouraged to take a license to the patent rather then attempt to fight or design around the patent. Each situation is an opportunity and requires a unique strategy. The following is one such example of strategic licensing.

Assume that Rattler Corp. has a patent on snake oil and has been selling the oil for several years at a rate of 10K units per month and 100% market share. When Cobra Inc., a Rattler competitor, begins selling a product that appears to infringe on Rattler's patent, Rattler's CEO wants to send Cobra a cease and desist order to get the company to stop making the product. Other alternatives include high royalties and strategic license pricing. The CEO believes that Cobra's entry will cost Rattler 30% of the market for snake oil.

With IP one must think strategically, not tactically or reactively. The goal is to make more money. To that end, due diligence and planning are the keys. Stopping the competition may not be the road to higher profits.

Assume that Rattler sells snake oil and other products through mainstream catalogs and major retailers. Cobra sells in specialty stores and in high end catalogs. In order to support its high prices, Cobra's advertising budget is ten times that of Rattler's budget.

Rattler's VP of Sales believes that Cobra will reach different customers through its high-end specialty channel. In her worst case estimation, Cobra would take 10% of Rattler's share of snake oil sales.

Rattler's VP of Marketing believes that Cobra's advertising will grow the total market and encourage Rattler's sales as the lower cost alternative. The marketing VP estimates that Cobra's presence may actually grow Rattler's sales by as much as 20%. Additionally, because Cobra sells high-priced products, the company has never mentioned the competition in its advertisements. Cobra appears to adhere to the marketing rule that mentioning the down-market competitors gives the low cost alternatives credibility.

One strategy is to wait for Cobra to get well-entrenched with its product and marketing. Afterward, Rattler should nicely approach Cobra regarding the patent issue and offer a reasonably priced license, which will more than make up for Rattler's potential loss of income. This strategy requires a forecast of the profits Rattler would have made on sales that Cobra may steal. The expected value estimating technique should be used. The potential increase in sales should be considered upside opportunity.

Among other terms, the licensing deal may require that Cobra advertise at a specified minimum level and must prevent Cobra from negatively portraying Rattler's brand.

In this strategy, the license price should be such that it is the best option for Cobra. It should be low enough that Cobra will not challenge or design around the patent. It should allow them to retain enough margin such that the snake oil remains an attractive product in its line. Furthermore, the price should be high enough to enhance Rattler's overall profitability.

Alternatively, expensive licensing terms or preventing Cobra from selling snake oil will elicit a different competitive response, which may include expensive legal battles regarding the injunction, expensive legal challenges to the patent, and poor relations with the competitor. Cobra may have IP which Rattler may be violating now or in the future. The relationship and licensing terms established in this particular negotiation are likely to carry forward into future negotiations.

RELATIONSHIPS AND LICENSING

Attempting to sell or license technology is a difficult task. Most potential licensors want potential licensees to sign non-disclosure agreements to protect the licensor from losing control of its intellectual property rights without appropriate compensation. Just the same, the potential licensee does not want to be precluded from continuing to develop technology it is already developing independent of the licensor. Major licensors and licensees understand these relationships; newcomers often do not. Many companies, especially in the toy industry, will only work with "professional inventors," which are companies and individuals with extensive experience closing licensing deals. As a result, cold-call selling of ideas is usually not effective.

Networking is the best way to sell and license intellectual property. Introductions through people who are trusted by both parties are typically a very effective way to increase the chances of a successful relationship. Organizations such as the Licensing Executive Society (LES) (www.les.org) exist to build and develop these network relationships.

In the world of intellectual property commercialization, it is "what you know and who you know" that will make the difference. Getting a foot in the door requires contacting the right person and having a relationship such that he or she will listen. Converting IP into CAISH requires that knowing precisely what you are selling and why they need to buy it.

CONCLUSION

The process of commercializing innovation starts with the idea, not the patent or the product. To maximize the return on investment, the value of an idea must be determined before the idea is transformed into an invention. In summary, the CAISH Model for Intellectual Property is:

Conceive It → Assess It → Invent It → Secure It →Harvest It.

When working with an existing patent or portfolio, the process is similar:

Define It → Assess It → Harvest It.

Companies and individuals alike make significant investments creating intellectual property and patents in particular. The process insures that the intellectual property is investment grade.

The process of studying, categorizing and managing the IP portfolio is generally ignored at most companies, and this ignorance results in missing huge opportunities to improve profits and generate a significant return on investment.

The intellectual property department should be a revenue center at major corporations. Converting IP to CAISH requires a team with a combination of business, technical and legal skills. The team must be focused on converting the non-strategic intellectual property into working assets through a process which consists of identifying high value intellectual property, finding licensing opportunities and closing deals.

TAKE AWAY

1. Patents must be thought of as business first, law second, and technology third.
2. It is estimated that only 4% of patents have any commercial value.
3. The CAISH process consists of Conceive, Assess, Invent, Secure, and Harvest.

4. Regardless of the path to invention, the success of the invention will be judged on its importance to the world or its ability to keep competitors out of the market.

5. If an idea or a solution can be protected through intellectual property such as patents or trademarks, then that specific feature can provide *ownable distinction*, where the feature is unique to a particular brand or a particular manufacturer.

6. Challenge the assumptions to see if the volume can be increased, the price can be increased, or the developmental costs can be reduced.

7. Be alert to the companies and industries that are developing and using the technologies. Such vigilance will identify potential licensing-in and licensing-out opportunities.

8. Rate of return, return on investment, and expected dollar value of profit, either for the entire project or for milestone options, can be used to compare the various opportunities.

9. Defining how the patent will be enforced will also affect the way the claims are crafted. Exercising a patent against a customer, government, global competitor or a charity may be difficult legally and politically as well as very costly.

10. Valuation of intellectual property is calculated using a blend of the three accepted methods: cost, income and market.

11. Many factors must be negotiated in a patent, trademark, trade secret or know-how licensing agreement. Such negotiation is a complex process.

12. Many companies develop intellectual property to differentiate their products, create barriers to competition and sustain revenue growth.

13. Networking is the best way to sell and license intellectual property. Engaging in introductions through people who are trusted by both parties is typically a very effective way to increase the chances of a successful relationship.

14. Commercialization is about "what you know and who you know."

MANAGING INNOVATION

Praveen Gupta

Competitive pressures are increasing, and innovation remains a prized approach to gain competitive advantage. Given the critical importance of innovation in the 21st century, managing investment and maximizing the corresponding returns become key responsibilities of every corporate leadership team. Such activity requires providing the right environment, providing organizational structure and incentives, and implementing the appropriate controls. Numerous tools and approaches have been shared to assist leaders of an innovation initiative to manage this process effectively. Now it is time for action!

Practicing innovation is an explicitly strategic initiative for some companies, while for others this practice is an implicit one. The challenge with a strategy for innovation arises due to a lack of understanding of the process. Now that the process is understood, as described in earlier chapters, precisely strategizing and planning for its execution become possible. The performance of the innovation strategy can be measured by using measurements across the supply chain and accelerated based on opportunities for improvement in the supply chain. Until now, innovation has been an inefficient R&D activity with unmeasured performance and unpredictable outcomes (and even a negative correlation with its inputs or resources). With a clearly-defined framework in terms of resources, knowledge, play and imagination, one can define the innovation process, establish a strategy for growth through new products and services with a certain confidence, and measure its success.

In order to launch the innovation initiative, the map of innovation must be understood and adopted. Figure 14.1, Innovation Map, shows that innovation starts at the top, where the leader commits to sustained profitable growth rather than just "making money." Growth-driven leaders are interested in innovation. Sustained growth can be realized if a growth office is appointed and headed by an innovation leader to ensure continual growth through innovation.

Having assigned a clear responsibility for innovation, a good understanding of the innovation process helps in making progress. The four components of a good innovation process are resources, knowledge, play and imagination. Without resources, innovation is not possible. Innovation does require investment of financial and intellectual resources. Once the resources are deployed, ideas for innovation are managed and commercialized as necessary to achieve a significant return on investment. The process continues from commercialization back to concepts to sustain and accelerate innovation.

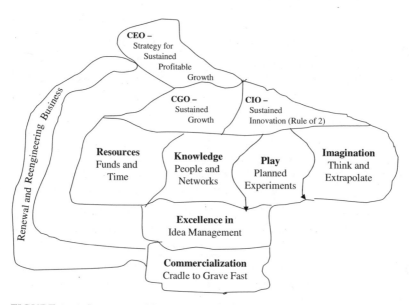

FIGURE 14.1 Innovation Map

RESOURCES FOR INNOVATION

Resources to support the commitment to innovation are a critical first action in order to execute the strategy well. Resources can either be financial or human capital. Proper utilization of human capital offers many opportunities for accelerating the innovation of new products and services. Of course, the human capital is an outcome of the financial investment in employees. In order to grow human capital, a corporation must develop tactics to invest in employees through training, exposure and experience in various aspects of business.

Human capital consisting of intellectual capability is one of the least utilized resources. Since so much opportunity exists, any improvement in its usage will have an amazing impact on the outcome. If the human brain's average utilization doubled from about 5% to 10%, the world would be a very different place. In such a scenario, a continual stream of new products or services would result. Innovation on demand and mass customization for customers would become the new status quo. When a switch is made from primarily utilizing mechanical resources to primarily utilizing intellectual resources, then nothing is standard anymore; that is the impact of such a change.

The financial capital still plays a role in providing support to the innovation strategy. Organizations must invest in facility and policy upgrades to create ambiance for creativity, an innovation room or physical space for experiments, and enough time to reflect and dig deeper into available mental resources. Organizations like to hire best-in-class people in all functions and keep them busy running around. People do not feel comfortable sitting in their office thinking because someone, including a supervisor, may think they are not doing any work. Managers love to see highly intelligent people run around, but they rarely allow them to use their brains.

One of the investments in innovation must be similar to 3M's allowance of 15% time for thinking, creativity, or learning without any accountability. Setting up an innovation room that provides facilities and resources for accelerated learning, such as a library with research material, knowledge management software, a laboratory for experimentation, resources for benchmarking, and a solitary space for deep thinking without any distraction, is an additional investment. Additional resources may be required for acquiring related new technology or tools to stretch organizational capability in new domains.

ORGANIZATION STRUCTURE

Interestingly, every organization has a CEO, COO and CFO, or their equivalents. Some organizations have a CTO, the Chief Technology Officer, for leading the new product development. Otherwise, most of the executive's attention is given to managing shareholders' perceptions of the organization as well as operational performance through cost reduction by pressuring suppliers and counting beans. Too much focus is on profit and not enough focus is on growth. Money is spent in cutting costs—sometimes more than what is spent on developing new products and services. Employees' jobs in the longer term are dependent on future opportunities, while still maintaining excellence in the portfolio.

In a typical R&D function, most of the resources are spent on development and little on research. As a result, the product performance is marginal and manufacturability is highly questionable. We quickly release new designs from the Design Engineering department to the Product Development department and wave goodbye. The Manufacturing department pays for the questionable designs due to lack of research and thoroughness in the design phase.

One of the reasons for marginal performance in manufacturing is attributed to the lack of defined targets for various parameters. Excellence in manufacturing is impossible without specifying targets for various parameters. In the absence of targets, the product is built to specification limits, and thus the product is produced with marginal performance.

Thus, organizations must be realigned for innovation and excellence. Without excellence, innovations will be wasted in marginal environments for performance. With excellence in operations, innovation can be accelerated because of faster and precise feedback from experiments. To create a structure for launching and sustaining the innovation initiative, organizations must have an innovation office or leader who can clearly bring economic sense to the design function and establish goals and processes to maximize use of human capital.

All innovation begins with an idea in some mind; thus excellence in idea management is a first critical step to building an innovative organization. The management process must be standardized, where every employee must contribute ideas to create value at the activity, process or product level. Starting a goal of one idea per quarter per employee is a reasonable goal upon which to build. Eventually, a continual flow of employee ideas towards development of new products or services in response to demand from customers or the marketplace should become the standard.

COMMUNICATING THE INNOVATION MESSAGE

Another important aspect in effectively executing the innovation strategy is the story. Leadership must develop a coherent message demonstrating the need for continual innovation, the benefits of innovation, and the consequences for not innovating enough to keep up with the customer demand. Consistency and constancy of the message are critical in generating employee interest across the organization, minimizing conflicts, and aligning organizational resources towards the common goal. Typically, the story may include a finding from the benchmarking study identifying opportunities to gain market share and improve, or even create a message to achieve fundamental business objectives.

In many companies, the innovation message is subtle through actions of certain individuals in executive positions. For example, Steve Jobs at Apple drives and demands innovation. At Google the work environment, also known as Googleplex, is committed to innovation and improving the "great" performance that promotes innovation at Google. Googleplex presents a unique lobby and hallway décor, office clusters, recreation, a café, a snack room, and dogs to promote free thinking and, combined with the goal, to develop innovative search-related products that promote continual innovation. The famous "15% investment in unaccountable time for employees to use" is a great way to communicate commitment to innovation.

INCENTIVES AND CONTROLS

On a personal level, incentives have played a significant role in my desire to do things differently over the course of my career. The Silver Quill awards at Motorola given for publishing articles challenged me to write my first paper. Besides getting $100 per published page, the company also encouraged more writing by making the reward continual. In addition, filing any application for a patent at Motorola resulted in at least $300 in the early to mid-1980s, and the reward amount continued to grow as the application moved further down the patent filing process. In both cases, the reward was given for starting to do things differently.

Once the flavor of doing things differently is tasted, it becomes difficult to stop. Understanding the innovation process in the brain shows that a personal incentive for learning can be much more effective than simply the incentive of achieving a more innovative outcome. Without learning multiple subjects, accelerating innovation is difficult. Thus, some incentives for learning must be in place. Interestingly, in tough times learning incentives are the first to go at many companies, which only highlights operational problems leading to corporate troubles. Can anyone imagine turning around or growing a company without having learning employees and innovation in place? This scenario certainly sounds like a recipe for winding down business through continual cost-cutting and negative revenue growth.

CULTURE AND CHANGE

Culture and change are two nebulous aspects of a business. Culture is about how we interact with each other and how we make decisions. On a daily basis, these two acts depend on the corporate values. Thus defining the corporate values that can be preserved under dire circumstances is imperative. A methodology can become a strategy, a strategy can become the corporate values, and new values can become the culture. Thus, once the innovation strategy begins to be implemented, based on its success and widespread institutionalization, it can become a corporate value of doing things differently. Once the value is accepted and practiced by everyone, all employees simply do things differently. Thus, the innovation strategy can become part of the corporate culture.

One of the major topics within the realm of corporate leadership is "change." In one of his live seminars, Tom Peters asked a question, "How long does it take one person to change his mind?" He talked about managing resistance to change, how leadership gradually phases in new practices, employee resistance, and implementation that is fragmented due to resistance. Then, he answered his own question. He said the same people who resist at work actually change at home all the time when needed. When people are doing things in a certain way, and then asked to change their behavior, resistance will occur because of the unanswered question, "Why should one change?"

People change in no time if they know why they should adapt new practices, and how they would benefit from such changes. To impact the change, employees should see the benefits of innovation through incentives or recognition. In any case, Tom Peters' answer for time needed to change a mind is "a moment." Once the decision is made, change is made in the mind followed by the practice.

Figure 14.2, The Innovative Thinking Matrix, shows various aspects of an organization that impact the mind of an organization once the decision is made to institutionalize innovation. As reported earlier, an individual can change his/her mind any time; however, as a group of employees, different people choose different moments to make up their minds. If the leadership communicates the strategy, employee roles, expected practices, and desired outcome, more employees will make the decision to accept the innovation initiative faster. Then, the leadership must walk the talk and encourage innovation practices while making decisions.

Business Aspects	Conventional Thinking	Innovative Thinking
Purpose of business	Make money	Create value, and make money
Customer Demand	Satisfy	See as a larger opportunity
Leadership	Manage for quarterly profits	Lead to build a business
Decision Making	React to fix	Respond to solve systemically
Goal Setting	Easy to achieve in short term goals	Challenging long term goals
Market Analysis	Limited external knowledge	Extensive benchmarking
Direction	Random and personal	Driven by vision and values
Profitable Growth	Profit or growth	Optimized profit and growth
Organizational Values	Competitive and negative	Collaborative and positive
Employee Learning	Hire and stale skilled employees	Build and renew employee skills
Innovation	Flash of a genius	Learned skill
Improvement	Incremental	Aggressive
Method of Innovation	Brain Storming	Well defined process
Innovators	Selected few	Everyone
Resources for Innovation	Allocated sporadically	Invested continually
Building Block of Innovation	Clusters of people	Networked Individual

FIGURE 14.2 Innovative Thinking

The aspects, as listed in the Figure 14.2, range from purpose of the business, customer demand, decision making, organization values, employee learning, innovation method, and resources. The matrix highlights elements of the innovation process at the methodology and its context levels. For each aspect, examine the conventional thinking and the innovative thinking, and plan to move towards innovative thinking through policy, procedures and practice.

For example, the main purpose of business is to make money. One can make money in two ways—legal or illegal. Thus, the purpose of a business must be clearly stated. One good definition of this purpose is "to provide value to customers by doing the right things efficiently and making money." Customers pay for the value. A business cannot make money unless the customer pays. The market capitalization of the business depends upon the long-term execution of the strategy. Any short-term manipulation of the stock through some short-term intent can at best be considered manipulation. Such manipulation is not the purpose of a business.

One of the leadership decisions that drives innovation in an organization is to set aggressive goals for improvement and continual change or renewal. The definition of "aggressive" can be understood as the amount or rate of change that forces one to think and do differently. For example, if one decides to make 10% more money at the personal level, one would think certain incremental activities, either work hours or a bonus, caused the 10% gain. However, if one decides to earn 50% more than the previous years, one starts thinking seriously and asks questions like "What else can I do?" The level of change that forces us to "do differently" beyond our comfort level is labeled aggressive. Aggressive, as some say right away, is not a pie in the sky. Aggressive means stretching current capability, resources and thinking. Figure 14.2 can be used to assess corporate culture for the required changes in an organization.

IDENTIFYING GAPS

Launching the innovative initiative begins with an understanding of current practices. The objective is to identify strengths and weaknesses, build on strengths and build in the areas of weaknesses. This step leads to an action plan that can enable an organization to make progress. Many organizations already have similar diagnostics matrices or assessment tools. However, their adequacy is sometimes questioned because of the lack of a framework for the innovation process.

Having defined the innovation process in earlier chapters, assessing and establishing a baseline for an organization for elements of the innovation process is easier. Thus the assessment includes questions about strategy, leadership, process inputs, process activities, process outputs and measures of innovation. At the early stage, one needs to highlight critical areas for change in order to realize an innovation friendly organization.

To evaluate each aspect of an organization, one can simply assign a percent score based on the applicable approach, deployment, and results (see Figure 14.3). For example, in assessing strategic commitment for growth through innovation,

Item#	Aspects of Innovation	Score (%)
1	A strategic commitment has been made to drive growth through innovation.	
2	An executive has been assigned full time to lead innovation.	
3	A strategy has been executed to accelerate innovation.	
4	Sufficient resources have been committed to support innovation activities.	
5	Departmental goals have been established to develop innovative solutions at the process level.	
6	Leadership has established a prestigious award for an innovative solution that creates exceptional value.	
7	Leadership understands the innovation process, and actively promotes risk-taking and doing things differently.	
8	A process has been established to achieve excellence in managing employee ideas.	
9	All employees have been given access to the Internet for conducting research in real time.	
10	Employees are encouraged to rotate among various departments.	
11	Company has in-house library of industry and related books and journals, and has access to on-line research services.	
12	Continual learning is rewarded at all levels, and time is allowed for learning.	
13	There is a facility for employees to brainstorm, play or experiment to test their ideas.	
14	Employees are encouraged to 'think' for new ideas for improving processes, products and services.	
15	Measures related to CEO recognition, employee ideas, and revenue from new offerings have been established.	
16	Employees are free to give funny ideas, and are not afraid of failures.	
	Average =	
Legend	0 - 20 = Ad Hoc; 21 - 40 = Marginal; 41 - 60 = Practiced; 61 – 80 = Standardized; 81 – 100 = Proven	

FIGURE 14.3 Diagnostics of Innovation

one can look for clearly stated and documented objectives, its institutionalization through tactics and processes, and outcomes in terms of continual leadership interest in growth through incomes. Considering these three elements with equal significance for evaluating each statement, one can assign a percent score. As for grading guidelines, one can consider score of 0–20 as ad hoc, 21–40 as marginal, 41–60 as practiced sporadically, 61–80 for standardizing the practice, and 81–100 for achieving desired results. While assessing the organization's performance, one does not need to split hairs about the absolute score. The objective is rather relative significance in order to initiate some actions to start making progress. For benchmarking purposes, the overall average can be calculated for assessing future progress.

For many years, leaders wanted to have an initiative, such as "Make Money and Have Fun." Once the initiative was implemented, measuring how much money was made was easier, but measuring how much fun employees had while working in the organization was impossible. While creating a process for developing innovative ideas faster, the author realized that as people think in terms of good, crazy, stupid, and funny ideas (as described in earlier chapters), first it took longer, and then ideas became more innovative. Most importantly, a measure of having fun evolved from the process.

Thus, one way to know when employees are having fun while working at a corporation is to measure how many funny ideas are coming out from employees, or how freely employees can present funny ideas without fear. When employees are having fun, they can pretty much say whatever they want to improve the company performance; no idea is discouraged. Being an innovative organization, we need all the funny ideas employees can come up with to improve or develop products or services.

INNOVATIVE LEADERSHIP

The success of a new strategy, without questioning its formulation, depends upon how passionately the leader champions for its success. Innovation has been used either as a corporate "value" or a strategy to facilitate turnaround. In either case, the CEO or executive must believe in its intended outcome, successfully drive the organization by providing direction, resources and support, and continually engaging employees through timely feedback and follow up.

In many organizations, the leader focuses on profit and initiates cost-cutting measures, which may be necessary in the short term for a struggling company; however, in doing so the leader acts counter to innovative thinking. Success begins with a thought in the leaders' mind and is achieved through the leadership traits. Figure 14.4 organizes various leadership traits according to the process of innovation and lists corresponding approaches of an innovative leader. People do what their leaders do—not what they ask for. Successful leaders demonstrate these behaviors and thus set an example for others to follow.

Leadership Traits	Innovative Leader
Learning	Reads a lot about a variety of subjects; interacts with community groups, employees, customers, and suppliers
Listening	Listens well to all ideas for noise, and noise for ideas
Personal style	Takes risks and executes tasks well
Interaction with employees	Encourages doing things differently better
Interaction with customers	Listens to their needs, and accepts challenges
Interaction with suppliers	Demands partnership for innovative solutions
Interaction with shareholders	Seeks support for longer term performance
Giving feedback	Rewards successes, understands failures, and encourages experiments
Behavior	Presents himself as positively enthusiastic, energetic, and an exemplary person

FIGURE 14.4 Innovative Leadership

MAKING AN INNOVATION STRATEGY WORK

Lawrence G. Hrebiniak, in his book *Making Strategy Work*, provides a template for leading effective execution and change. According to Hrebiniak, and the methodology used for Six Sigma projects, one can address the following tactics for successful execution of the innovation strategy:

- Define a clear charter with cost and benefit analysis
- Identify stakeholders and utilize their influence
- Align organizational structure
- Develop a roadmap with clearly defined accountability
- Coordinate tasks and frequently share information
- Support and reinforce execution
- Manage change and culture
- Establish a process for sustaining innovation
- Reward success and inspire excellence
- Learn and adjust the strategy

A lot has been written about strategy execution; however, success depends on this ultimate factor: the desire of the leadership to make the strategy work. If a leader is committed to making innovation become an integral part of doing business, it will happen. Otherwise innovation will not happen.

IN CLOSING

"Business Innovation in the 21st Century" equips us with a better understanding of the innovation process. In the absence of a clear understanding of the innovation process, leadership starts the innovation initiative, commits resources, establishes measures, and then finds that innovation does not happen. However, with a clear understanding of innovation, a clearly defined strategy can be formulated and executed for expected outcomes.

The author has attempted to present a comprehensive approach to the innovation process through the science of innovation based on his research and experience, and by sharing the expertise of various contributors. Just like any problem, innovation has been an unsolved puzzle. The book prepares readers to solve the innovation puzzle in their own way and enjoy the experience of being innovative.

TAKE AWAY

1. Innovation begins with leadership's commitment to sustained profitable growth (rather than just "making money").
2. Providing resources to support the commitment to innovation is a critical first action in order to execute the strategy well.
3. The human capital is an outcome of the financial investment in employees through training, exposure and experience in various aspects of business.
4. Organizations must invest in upgrading facilities and policies to create ambiance for creativity (such as an innovation room or physical space for experiments) as well as allowing time to reflect and dig deeper into the available mental resources.

5. One of the investments in innovation must be similar to 3M's allowance of 15% time without any accountability for thinking, creativity or learning.

6. To create a structure for launching and sustaining the innovation initiative, organizations must have an innovation office or leader who can clearly bring economic sense to the design function and establish goals and processes to maximize the use of human capital.

7. Leadership must develop a coherent message demonstrating the need for continual innovation, benefits of innovation, and consequences for not innovating enough to keep up with the customer demand.

8. Without continual learning, accelerating innovation will be difficult. Thus some incentives for learning must be present.

9. In tough times learning incentives are the first to go, which only highlights operational problems leading to corporate troubles.

10. People change in no time if they know why they should adopt new practices and how they will benefit from such changes.

11. When employees give funny ideas freely without fear of criticism, the organization's innovation culture is pretty apparent. Funny ideas are innovative ideas.

12. The CEO or leadership must believe in the outcome of innovation and drive it through the organization by providing direction, resources and support, and by continually engaging with employees through timely feedback and follow-up.

13. People do what their leaders do—not necessarily what those leaders request of them. Successful leaders demonstrate innovative practices and thus set an example for others to follow.

14. Deploying employee intellectual capability will grow intellectual power and produce more innovative products and services.

FINAL THOUGHTS: WISDOM OF INNOVATION

Robert W. Galvin

Robert Galvin, the son of Motorola Inc. founder Paul Galvin, started working for the company as a teenager in 1940. By 1959, he was the CEO of the company and served as chairman until his retirement in 1990. Under his guidance, Motorola became one of the world's largest cell phone manufacturers. Galvin has remained dedicated to the technology, and underlying science behind it, by actively supporting the organizations that research new technologies.

"Over a 60-year career, Bob Galvin has had a profound effect on economic well-being, technological innovation, telecommunications and public service in the United States and around the world," says David Theroux, founder and president of the Independent Institute. "Future business leaders would be well-served to look to him as a role model."

In this context, the author conducted this interview to capture Bob Galvin's wisdom of innovation. Interestingly, Bob has abridged Alex Brown's book *Your Creative Power*, which has a recipe for bringing out the best from people. Bob says the whole book could be summarized on a page. A summary of Alex Osborn's book is included at the end of the interview.

The author would like to thank Bob Galvin for his generosity of precious time and priceless secrets of innovation, which he exemplified while leading Motorola to success.

THOUGHTS ABOUT CREATIVITY AND INNOVATION

In the 1940s Alex Osborn activated a process called brainstorming, which sounds superficial. Intellectuals, who think the brilliant look for degrees, thought brainstorming was for ordinary people. I am looking for the other process of thinking. As youngsters in high school, we have mental capability such that all was towards reason, logic, deduction and induction. Every time teachers asked us to write a creative theme, I asked the teacher(s) how I could have creative thoughts. I never got the answer!

In college, similarly we were supposed to have creative ideas. I was of the mind of having a factor or a method to find an answer to the question of how to get creative ideas. I was continually alert to something and/or someone providing an answer. I found Alex Osborn's book was a formula for creative ideas.

To me creativity is one of the two fundamental intellectual exercises in which the mind can engage (a set of steps according to Osborn.) I memorized those steps. Osborn's theory and process I practiced in experiments in my 20's. I had the 'allowability' due to being the 'boss's' son. People tolerated it because my ideas were relevant. I used them daily.

Random is a superb word. We should be able to organize in random ways. It is important to engage in a timely fashion to think counter-intuitively. Most people find it lazily easy to have a comfortable instinct. We are biased towards our instincts. It was important when I witnessed other peoples' subjects. When a subject seems to be flowing, I automatically raised the question quietly, respectfully challenging the flow. I wanted people to look at the vector substantially differently. You cannot come up with a good solution unless you look at all choices. Thinkers are not considering axiomatic thinking when a flow of thoughts is established. Insufficiency and superficiality of thinking settles in without knowing 'what we are missing." I wonder if they have thought of all the appropriate factors.

Osborn inculcated in me the fundamental steps of creativity because they were very valuable. I would use them four to five times a day in pretty ordinary settings. Once a Senior Executive at Motorola, who would not compliment me for personal reasons, made a comment that I thought was quite flattering. He said, "How do you get so many good ideas?" I replied, "Getting ideas is a combination of intellectual engagement and the Osborn process you just do all the time. Sometimes a designated place helps. I suppose there is a place for a place. However, Creativity or Innovation should be a continual disposition. Creativity can occur anywhere and everywhere. In one engagement we went to breakout rooms for augmenting the creativity process. However, it would be wrong to tell somebody to go to a 'place' just for innovation.

There was a time, for the benefit of all employees, that we paid for innovation. Incentives can be a founding traction for innovation. I myself did not come up with many ideas that were great. Instead, I used my office to be the place where creative people would get together in the evening, practice an intellectual process to display every body's ideas, and come up with a better combination of ideas. People would draw something, kept staring at it, and suddenly someone would come up with an alternative. Instead of juggling ideas in our minds, we would do it collectively on a board or something like that. We experimented almost daily.

About thirty years ago, we had the privilege to take our children on a trip to the Greek Islands. We had a big yacht and a professor as a guide from which to learn history. We rode a boat down to Turkey on a Sunday and went to a museum—a small crummy-looking one! We found somebody to open it, as we thought it might have some tertiary value. There we saw two tables. One had goblets from the time of Christ, around 2000 years old. The other table had goblets about 1000 years old. I saw them as two points in time. I connected them and drew a mental line. There we witnessed a very visual change in technology of over 1000 years. One set of goblets were looking rough and without polish; the other set looked a little shinier and polished with some designs. They both were made from sand!

I brought the kids back in the room and said to the children, "We cook glasses better than those 'glasses.'" In other words, we drew a line in the sand and could show the rate of change of wisdom over 1000 years. In the first thousand years, we did little better from rough to polished glasses. Over the next 1000 years, glasses became beautiful in design, finish and colors. I thought we could do something else with the sand. We could do a better job; we could make sand work for us. Rate of change is a major change the civilization is going through today. My mental process was able to put the whole process together because of my disposition. I mean we were talking about change in sand over two thousand years and its marvels. We make chips today with sand. I believe kids think today even faster than I could think then.

Ultimately, one would like to accomplish with a large number of people who feel no stress, and who have memorized a set of practices and processes such that every time they are provoked, they think about it. One of the things a person must adequately possess is confidence, which is different from arrogance, in dealing with various subjects. Meeting in my office for about 30 days with my people, I came up with four or five stupid ideas, just to encourage others to come up with better ideas with self-confidence.

CREATING AN INNOVATIVE ENVIRONMENT

The credit for creating an environment for innovation at Motorola sublimely goes to my father, Paul Galvin. My father had a brilliant realization that whatever he started with had to be renewed. The paramount word is renewal. My father had reinforced the significance of it. All, my father wanted to do was to be in business. He had no idea of market, research, development, and 'innovation.' He was in business when he was 11 years old. He knew that he had to have something new every two years, and he did it a few times. Then I came along and found the word 'renewing.' Renewal is the driving thrust of institutional growth. My father invoked this sense in me. We did promulgate renewal in our speeches. We publicized and recited, **"The driving thrust of the institution is renewal."** Whether at the activity or the thought level, 'renewal is a driving thrust' must become an attitude.

How one communicates or manifests a message is also important. Packaging or putting a thought in telegraphic picturesque language is important for people to embrace the message. Leaders who are listening to their respective associates, there is nothing that speaks more than hearing a comment, "Let's try this," or for the boss to say, "Why not?" instead of saying "Why would you do that?" A challenge to stifle the idea would be imprudent. With people whom I trusted (and I trusted many of them), if they said, "Bob, let's try this" or "Let's try that," I said, "**WHY NOT!**"

Another factor that made me comfortable was when I heard "Let's start" vs. "Let's think about it," (i.e., thus preventing an idea, or worrying from failures.) All of us are capable of juggling many things and trying crazy ideas. Many times, as a new idea is proposed, managers think of the total cost and risk associated with it. They defer the decision to try it out. One must understand that the cost of trying an idea is incremental. Therefore, trying out a new idea can be managed before we drive the ship over to the cliff. Iridium was the worst approval I gave which cost us a lot of money over the years. We all lose sometimes and learn humility. Interestingly, I was at a meeting recently with a few loyalists. Someone mentioned how successful the Iridium has been for gathering intelligence recently. An idea starts with a small investment which grows with development over time. It was far better that we tried and did not make it successful, than not even trying it at all. Sometimes, you just never know unless you try it out.

The authorities can daily practice a picturesque culture, visual communication and renewal by encouraging new ideas by saying "Why not," "Let's try it," or "Let's start." Typically, a middle boss would not trust others and holds an idea back until the idea is proven. Lots of things happened that we played secretly. Having some projects secretly done was perfectly fine for me. Of course, we would know about any big ones, just in case.

Finally, our intellectual processes have to have fuel through the network. Employees can participate in networks where they get a chance to renew their thinking, get new exposures, and continue to be engaged intellectually in developing new ideas. That is the idea of ideas—the innovation!

Nuggets from *Your Creative Power* **by Bob Galvin, Motorola University Press, 1991**

(An Abridged Version of the original book Alex Osborn by Alex Osborn (www.osborn.org)

1. Creative power can be stepped up by **effort**; there are ways we can guide our creative thinking. With enough creative effort, **each of us** could find the ideas that would smooth out our rocky roads.

2. What we need is a conscious appreciation of the fact that ideas have been, and can be, the **solution to almost every human problem**. All of us possess this talent. Most of us have more imagination than we ever put to use. It is often latent—brought out only by **internal drive** or by force of circumstances.

3. The point is that the degree of one's creative power does not depend upon a degree. This point is stressed because **self-confidence** is one of the keys to increased creativity.

4. Even if our native talent should stop growing when our body stops growing, it would still be true that our creative ability can keep growing year after year in pace with the effort we put into it.

5. If we **set aside a definite period** for creative thinking we can best lure the muse.

6. With **proper concentration** it is possible to track down ideas anywhere, at any time. Concentration is nothing but attention, sharply focused and steadily sustained. It is an acquired habit rather than a native gift.

7. The creativity is more than mere imagination. It is imagination inseparably coupled with both intent and effort.

8. When thinking creatively in groups, **association is a powerful factor**. We bat ideas around the table and one idea bumps another into existence.

9. **Emotional drive** is self-starting and largely automatic, whether based on hunger, fear, love or ambition.

10. **Note-taking** helps in several ways. It empowers association, it piles up alternatives, and it stores rich fuel that

otherwise would trickle out through our forgettery. But, above all, note-taking itself induces a spirit of effort.

11. In the average person, judgment grows automatically with years, while creativity dwindles unless consciously kept up.

12. In getting going, keeping going, or giving out, we have every person to **sweep timidity aside** and gird our efforts with **courage.**

13. The best policy is **always to keep suggesting.** You may develop a reputation as a crackpot, but as soon as one or two of your ideas materialize, your employer and co-workers begin to give you serious consideration.

14. For all of us, a good rule is always to **encourage ideas**— to encourage speaking up as well as thinking up.

15. Actual doing is, of course, the best exercise. The way to create is to create, just as the way to write is to write.

16. Even open-minded people may have to ward off influences that could close their minds while in quest of an idea.

17. Before we set our aim, let's flex ourselves, open our minds, intensify our intent, court awareness, encourage curiosity and then tug that boot-strap marked "concentration." Thus we can get into a working mood where effort is more like a sport.

18. One idea leads to another; one aim often leads to another.

19. Analysis of any kind can of itself be creative 'fruit,' for it tends to uncover clues which speed up our power of association and thus feeds our imagination.

20. There are two kinds of specific facts we should seek— those which are **inherent** in our problem and those which may have **some bearing.**

21. The basic principle is variation. The active adjunct to the principle of variation is plenty of alternatives. We need to pile up **alternatives.**

22. Quantify, quantify, and more quantify! This is the surest recipe for ideas.

23. **Exaggeration** is but one of the many byways which lead off from the magnification highway. By sending our imagination

down these trails, we can add more alternatives; the more numerous the alternatives, the better the ideas. In turn, the conscious effort we put into such quests tends to step up our creative power.

24. The **substitution** train is an endless road to an infinite number of ideas.

25. There are so many little ways in which we can work via **vice versa** in our relations with each other.

26. **Unexpected kindliness** can do wonders in business.

27. Illumination is short snatches of unconsciousness work, or the intellectual rhythm of ideas. Illumination, which remains a mystery like life itself, can also be coaxed by **shifting our minds to another subject**.

28. Illumination comes while coasting, but coasting inescapably implies that power has been previously applied. A tragic tendency of mental Micawbers is to overrate illumination and underrate effort. The fact is that the ideas we receive while idling are quite often by the way of extra dividends.

29. Accidents are seldom the answers. Good breaks count most in what they lead to—if we **follow through**. Sometimes, we fail to grow the seed of an idea.

30. When stumped in the course of a creative project, we need to stop and review. We should analyze the problem anew, think up **still other alternatives**, and then proceed all over again.

31. Despite advances in organized research, the **creative power of the individual** is what still counts the most.

32. The spirit of a brainstorm session can make or break it. **Self-encouragement** is needed almost as much as **mutual encouragement**.

33. To induce creativity, educators should do their best to **arouse enthusiasm for imaginative thinking**, encourage every creative effort on the part of their pupils, and act as creative coaches.

34. Every business, big or little, needs spark plugs—lenders who have ideas and know how to make them click.

35. Not only in business but in every line, the **quality of leadership depends on creative power.**

BIBLIOGRAPHY/ REFERENCES

CHAPTER ONE

1. http://www.virtualclassroom.net/tvc/internet/fire.htm
2. http://neon.mems.cmu.edu/cramb/Processing/history.html
3. http://www.c3.hu/scca/butterfly/Kunzel/synopsis.html
4. http://userpage.fu-berlin.de/~rober/linguistics/origins.html
5. http://college3.nytimes.com/guests/articles/2003/07/15/1100994.xml
6. http://www-groups.dcs.st-and.ac.uk/~history/Mathematicians/Galileo.html
7. http://galileo.rice.edu/chron/galileo.html
8. http://web.class.ufl.edu/users/rhatch/pages/13-NDFE/newton/05-newton-timeline-m.html
9. http://www.aip.org/history/einstein/index.html
10. http://www.hfmgv.org/exhibits/hf/default.asp
11. http://www.lucidcafe.com/library/96feb/edison.html
12. http://www.thomasedison.com/biog.htm
13. http://www.oecd.org/document/28/0,2340,en_2649_34273_34243548_1_1_1_1,00.html
14. http://www.idrc.ca/en/ev-33214-201-1-DO_TOPIC.html
15. http://trendchart.cordis.lu/scoreboards/scoreboard2003/index.cfm
16. http://www.usembassy-china.org.cn/sandt/stconfaug99.html
17. http://topics.developmentgateway.org/knowledge/rc/BrowseContent.do~source=RCContentUser~folderId=3212
18. http://www.indianexpress.com/full_story.php?content_id=69486
19. www.compete.org
20. Utterback, James M., Mastering the Dynamics of Innovation, HBS, MA 1996
21. Porter, Michael E., Clusters of Innovation Initiative, Final Report, Council of Competitiveness, http://www.compete.org/pdf/pitts_final.pdf

CHAPTER TWO

1. Conger, J. A., 1995. Boogie down wonderland: Creativity and visionary leadership. In C. M. Ford and D. A. Gioia (Eds), *Creative action in organizations*. Thousand Oaks, CA: Sage.
2. Csikszentmihalyi, M., 1996. *Creativity: Flow and the psychology of discovery and invention*. New York: Harper Collins.

CHAPTER THREE

1. Harrington, H. James, *The Creativity Toolkit: Provoking Creativity in Individuals and Organizations*, McGraw Hill, NY 1997

CHAPTER FOUR

1. "Business delegates praise federal innovation strategy, but Kyoto rankles." The Canadian Press, September 24, 2002, 12:24 am.
2. "CEBIT Schroeder launches innovation strategy to complement structural reform." AFX International Focus, March 17, 2004.
3. Christensen, C. "The innovator's dilemma."
4. "Defra's science and innovation strategy published." *Government News Network, May 30, 2003.*
5. "Genome BC funding boosts innovation strategy." *Elsevier Engineering Information, March 19, 2003.* Press Release (Chemicals).
6. "Government of Canada launches innovation strategy." *Canada NewsWire, February 12, 2002.*
7. Henderson, R.M. & Clark, K.B. "Architectural innovation: The reconfiguration of existing product technologies and the failure of established firms (technology, organizations, and innovation)." *Administrative Science Quarterly 9, March 1, 1990.*
8. Hoppe, H.C. & Lehmann-Grube, U. "Innovation timing games: A general framework with applications." *Journal of Economic Theory—30 Vol. 121:1, March 1, 2005.* ISSN: 00220531.
9. "Innovation strategy needs action (Editor's Notebook)." *Canadian Machinery and Metalworking 3, Vol. 97:6, July 1, 2002.* ISSN: 0008-4379.
10. Loudon, A. "Webs of innovation: the networked economy demands new ways to innovate." FT.com series of Financial Times *Books for the Future Minded*, Pearson Education. ISBN 0273656465.
11. Mellish, M., & Allen, L. "Drug makers swallow bitter pills." *Companies and Markets*, Australian Financial Review, July 20, 2002.
12. Meyer, C. VP of Cap Gemini, Ernst and Young and Director of its Center for Business Innovation in Cambridge. "HBR August 2002."
13. "NE Ohio should create innovation strategy." *Crain's Cleveland Business 8, Vol. 25:29, July 19, 2004.*
14. Puhlmann, M., & Gouy, M. "Internal barriers to innovation—An international investigation." *Pharmaceutical Executive—84 Vol. 19:6; June 1, 1999.* ISSN: 0279-6570
15. Quinn, J.B., Baruch, J.J., & Zien, K.A. "Software-based innovation." *The McKinsey Quarterly, September 22, 1996.*
16. Rachlis, M. "Medicare made easy: The solution to our health-care funding problem is innovation." *The Globe and Mail, April 26, 2004.*
17. Sebastian, P. "A special background report on trends in industry and finance." The *Wall Street Journal, September 13, 1990, p. A1.*

CHAPTER FIVE

1. *"Built to Last—Successful Habits of Visionary Companies"*, by James C. Collins & Jerry I. Porras
2. *"Six Thinking Hats"* by Edward De Bono
3. *"The S Curve"*, by Rajagopal Sukumar in Joe Kissell's Interesting Thing of the Day, September 25, 2004
4. www.yokoninnovation.ca

CHAPTER SIX

1. Bear, Mark F., Connors, Barry W., and Paradiso Michael A. Neuroscience: Exploring the Brain, 2nd Ed., Lippincott Williams and Wilkins, Baltimore, 2001.
2. Hawkins, Jeff and Blackeslee, Sandra. On Intelligence, Times Books, New York, 2004.
3. McEntarffer, Robert, and Wesely, Allyson J. How to Prepare for the AP Psychology Advanced Placement Examination, Barron's Educational Series, Hauppauge, 2004.
4. Purveys, William K., Sadava, David, Orians, Gordon H., Heller, Craig H. Life—The Science of Biology, 7th Ed., W. H. Freeman Company, Gordonsville, 2004.
5. Restak, Richard. Brainscapes, Hyperion, New York, 1995.
6. Wilson, Taylor Andrew. The Mind Accelerator, Volition Thought House.
7. http://www.benbest.com/science/anatmind/anatmind.html
8. http://faculty.washington.edu/chudler/baw1.html
9. http://www.human-evolution.org/human_psych101.php
10. http://www.getimusic.com/brain.php
11. http://www.waiting.com/brainanatomy.html

CHAPTER SEVEN

1. Altshuller, G. And Suddenly the Inventor Appeared: TRIZ, the Theory of Inventive Problem Solving. Technical Innovation Center, Worcester, MA, 1996.
2. Drucker, Peter F. The Discipline of Innovation. Harvard Business Review, August, 1985.
3. Drucker, Peter F. Innovation and Entrepreneurship: Practice and Principles. Harper & Row, 1985.
4. Edison Effect—http://www.ieee-virtualmuseum.org/collection/tech.php?id=2345876&lid=1
5. Gupta, Praveen. The Six Sigma Performance Handbook. McGraw-Hill Publishers, New York, 2004.
6. Gupta, Praveen. 4P's Cycle of Process Management. Quality Progress, April 2006.

7. McCarty, Thomas; Daniels, Lorraine; Bremer Michael; and Gupta, Praveen. The Six Sigma Black Belt Handbook. McGraw-Hill Publishers, New York, 2004.

CHAPTER EIGHT

1. 1999 HP Annual Report (as captured on www.creativityatwork.com)
2. www.stealcase.com
3. www.hbdi.com
4. Peter Geyer, INTP
5. www.enneagraminstitute.com
6. www.hbdi.com/diversity.html

CHAPTER NINE

1. Chasan, Emily. *CEOs find innovation hard to achieve—survey*, Innovathttp://prelaunch.reuters.com/sponsoredby/AMEX/article.asp, 2006
2. Christensen, Clayton M. and Raynor, Michael E. *The Innovator's Solution*. HBS Press, MA, 2003.
3. Gupta, Praveen. *Innovation and Six Sigma, Six Sigma Columns*. www.qualitydigest.com, December 2004.
4. Gupta, Praveen. *Innovation: The Key to a Successful Project*. Six Sigma Forum Magazine, August 2005.
5. Gupta, Praveen. *Six Sigma Business Scorecard: A Comprehensive Corporate Performance Scorecard*. McGraw-Hill, NY, 2003.
6. Gupta, Praveen. *The Six Sigma Performance Handbook*. McGraw-Hill, NY, 2004.
7. McCarty, Tom; Daniels, Lorraine; Bremer, Michael; Gupta, Praveen. *The Six Sigma Black Belt Handbook*, McGraw-Hill, NY, 2004.
8. http://www.eurescom.de/message/messageMar2003/International_Symposium_on_Innovation_Methodologies.asp
9. Edison Effect : http://www.bookrags.com/sciences/physics/electronics-wop.html

CHAPTER TEN

1. Basili, V.R. (1992). "Software modeling and measurements: The goal question metric paradigm." Computer Science Technical Report Series, CS-TR-2956 (UMIACS-TR-92-96). College Park: University of Maryland.
2. National Innovation Initiative, Innovate America Report, 2nd Edition, www.Compete.org, Washington, D.C. 2005.
3. National Innovation Initiative Summit and Report.

4. (http://compete.org/pdf/NII_EXEC_SUM.pdf), Washington D.C., 2005.
5. Neely, A. "In search of a metric system for innovation." Financial Times, October 7, 2004.
6. Zoghi, Cindy; Mohr, Robert D. and Meyer, Peter, "Workplace Organization and Innovation," Unpublished Work, U.S. Bureau of Labor Statistics, and University of New Hampshire, May 2005 (Used with permission).
7. Rogers, Mark. "The Definition and Measurements of Innovation." http://ecom.unimelb.edu.au/iaesrwww.home.html, The University of Melbourne, Australia, 1998.
8. Studt, Tim. "Measuring Innovation . . . Gauging Your Organization's Success." R&D Magazine (www.rdmag.com).
9. "The Climate for Creativity, Innovation and Change." The Creative Problem Solving Group, Inc., www.cspb.com.
10. Walcott, Robert P. "The Innovation Radar." Perspectives to Date, 2003, (http://www.technologymanagementchicago.org/meetings/presentations/03-10.pdf)
11. http://www.paconsulting.com/services/tech_innovation/innovation/entry_high_performance.htm.

CHAPTER ELEVEN

1. Andrew, James P. and Sirkin, Harold L. (2003), "Innovating for cash," Harvard Business Review, Vol. 81 No. 09, pp. 58–68.
2. Carlzon, Jan (1993), "Moments of Truth," 9th Brazilian edition "A Hora Da Verdade, publisher COP Editora, Rio de Janeiro, Brazil.
3. Chesbrough, Henry W. (2005), "Toward a new science of services," Harvard Business Review, Breakthrough Ideas for 2005, Vol. 83 No. 2, pp. 43–44.
4. Conley, Lucas (2005), "Cultural phenomenon: Umpqua Bank is changing the culture of customer service at banks," Fast Company magazine, April 2005, Issue 93, pp. 76–77.
5. Corporate Strategy Board (2001), "From services to solutions: Building the capabilities for customer-oriented strategies," Corporate Executive Board, 2001, Washighton, DC.
6. Davis, Scott (2002), "Implementing your BAM strategy: 11 steps to making your brand a more valuable business asset," Journal of Consumer Marketing, Vol. 19 No. 06, pp. 503–513.
7. Day, G. S. (1994a), "Continuous learning about markets," California Management Review, Vol. 36 No. 4, pp. 9–31.
8. Day, G. S. (1994b), "The capabilities of market-driven organizations," Journal of Marketing, Vol. 58 No. 4, pp. 37–52.
9. Eiglier P., Langeard E. (1991), "Servuction: le marketing des services," Portuguese edition "Servuction: a gestão marketing de empresas de serviços," publisher Editora McGraw-Hill de Portugal.

10. GE (2004), 2004 General Electric Letter to Stakeholders, pp. 1–11. URL:http://www.ge.com/files/usa/en/ar2004/pdfs/ge_ar2004_letter.pdf. Access date: 03/27/2005.
11. Goncalves, Alexis P. (1998), "Quality Management in South America," Quality Progress Magazine, ASQ, August 1998, Vol. 31, No. 8, pp. 124–126.
12. Goncalves, Alexis P. (2003), "Aligning Six Sigma to the Voice of the Customer," Six Sigma Summit 2003, IQPC, January 21–22, 2004, Miami, FL.
13. Goncalves, Alexis P. (2004), "Amplifying VOC through Innovation," Conference on Six Sigma for Financial Services: Impacting on the Customer Experience, IQPC, July 27–28, 2004, New York, NY.
14. Grönroos, Christian (1990), "Service management and marketing: managing the moments of truth in service competition," Lexington Books, New York, NY.
15. Gummesson, Evert (1988), "Service quality and product quality combined," Review of Business, Vol. 09 No. 03, pp. 14–19.
16. Gummesson, Evert (1994), "Service management: an evaluation and the future," International Journal of Service Industry Management, Vol. 05 No. 01, pp. 77–96.
17. Heskett, James L.; Jones, Thomas O.; Loveman, Gary W; Sasser, W. Earl; and Schlesinger, Leonard A. (1994), "Putting the service-profit chain to work," Harvard Business Review, March-April 1994, Vol. 72 No. 02, pp. 164–175.
18. Kok, R.A.W., Hillebrand, B. and Biemans, W.G. (2002), "Market-oriented product development as an organizational learning capability: findings from two cases," Research Report, University of Groningen, Research Institute SOM (Systems, Organizations and Management).
19. Levitt, Theodore (1972), "Production-line approach to service," Harvard Business Review, September-October 1972, Vol. 50 No. 05, pp. 91–102.
20. Levitt, Theodore (1976), "The industrialization of service," Harvard Business Review, September-October 1976, Vol. 54 No. 05, pp. 107–118.
21. Lovelock, Christopher H. (1983), "Classifying services to gain strategic market insights," Journal of Marketing, Vol. 47, Summer 1983, pp. 9–20.
22. Lovelock, Christopher H. and Yip, George S. (1996), "Developing global strategies for service business," California Management Review, Vol. 38 No. 02, pp. 64–86.
23. Lundkvist, Anders and Yakhlef, Ali (2004), "Customer involvement in new service development: a conversational approach," Managing Service Quality magazine, Vol. 14 No. 2/3, 2004, pp. 249–257.
24. McFarland, Jennifer (2001), "Margaret Mead meets consumer fieldwork: a prime on ethnographic market research," Harvard Management Update, a newsletter from HBS Publishing, August 2001.

25. McGregor, Jena (2004), "The World Is Their R&D Lab," Fast Company magazine, May 2004, Issue 82, pp. 35–36.

26. Moore, Johnnie (2005), "Barista Bankers," Johnnie Moore's Weblog, published February 7, 2005. URL: http://www.johnniemoore.com/blog/archives/000733.php. Access date: 03/20/2005.

27. NIST (2003), National Institute for Standards and Technology, Baldrige National Quality Program, 2003 Award Recipients Applications Summaries, Caterpillar Financial Services Corporation.

28. Normann, Richard (1991), "Service management: strategy and leadership in service business," 2nd edition, John Wiley & Son, London, England.

29. Peppers, Don; Rogers, Martha and Hornby, Richard (2004), "Customer intimacy in financial services," SASCOM online magazine, Second Quarter 2004.

30. Prahalad, C.K. and Ramaswamy, Venkatram (2002), "The co-creation connection," Strategy + Business magazine, Second Quarter 2002, Issue 27, pp. 01–12.

31. Reed, Will (2004), "Hypnotic Marketing," Guerrilla Marketing Genius newsletter, August 2004, pp. 1–2.

32. Reichheld, Frederick F. (2003), "The one number you need to grow," Harvard Business Review, December 2003, Vol. 81 No. 12, pp. 46–55.

33. Reichheld, Frederick F. and Teal, Thomas (1996), "The loyalty effect: The hidden force behind growth, profits, and lasting value," Harvard Business School Press, Boston, MA.

34. Reichheld, Frederick F. and Sasser, W. Earl (1990), "Zero defections: Quality comes to service," Harvard Business Review, September-October 1990, Vol. 68 No. 05, pp. 105–111.

35. Saco, Roberto M. (1997), "The criteria: A looking glass to American's understanding of quality," Quality Progress Magazine, ASQ, November 1997, Vol. 30 No. 11, pp. 89–96.

36. Salter, Chuck (1998), "Progressive makes big claims," Fast Company magazine, November 1998, Issue 19, pp. 176–178.

37. Salter, Chuck (2002), "Service: Commerce Bank," Fast Company magazine, May 2002, Issue 58, pp. 80–88.

38. Schumpeter, Joseph A. (1934), "The theory of economic development: An inquiry into profits, capital, credit, interest and the business cycle," Oxford University Press, 1961, New York, NY.

39. Senge, Peter M. et al. (1994), "The Fifth discipline fieldbook: Strategies and tools for building a learning organization," Doubleday Publishing, New York, NY.

40. Seybold, Patricia B.; Marshak, Ronni T. and Lewis, Jeffrey M. (2001), "The customer revolution," Crown Business, New York, NY.

41. Sharma, Deven; Lucier, Chuck; Molloy, Richard (2002), "From solutions to symbiosis: Blending with your customers," Strategy + Business magazine, Second Quarter 2002, Issue 27, pp. 13–23.

42. Shostack, G. Lynn (1984), "Designing services that deliver," Harvard Business Review, January-February 1984, Vol. 62 No. 01, pp. 133–139.
43. Sundbo, J. (1994), "Modulization of service production and a thesis of convergence between service and manufacturing organizations," Scandinavian Journal of Management, Vol. 10 No. 3, pp. 245–266.
44. Swan, Kate (2003), "Bank of (Middle) America," Fast Company magazine, March 2003, Issue 68, pp. 104–106.
45. Swank, Cynthia Karen (2003), "The lean service machine," Harvard Business Review, Vol. 81 No. 10, pp. 58–68.
46. Tapscott, Don (1996), "The digital economy: Promise and peril in the age of networked intelligence," McGraw-Hill, New York, NY.
47. Thomke, Stefan (2003), "R&D comes to services: Bank of America's path breaking experiments," Harvard Business Review, Vol. 81 No. 4, pp. 71–9.
48. Von Hippel, Eric and Sonnack, Mary (1999), "Breakthroughs to Order at 3M," MIT-SSM Working Paper, January, 1999.
49. Von Hippel, Eric and Thomke, Stefan (2002), "Customers as innovators: A new way to create value," Harvard Business Review, Vol. 80 No. 4, pp. 35–45.
50. Von Hippel, Eric (2005), "Democratizing Innovation," The MIT Press, Cambridge, MA.
51. Walden, David et al. (1993), "Kano's methods for understanding customer-defined quality," Center for Quality of Management Journal, Fall 1993, Vol. 02 No. 04, pp. 03–28.
52. Womack, James P. and Jones, Daniel T. (2005), "Lean consumption," Harvard Business Review, Vol. 83 No. 3, pp. 58–68.
53. Zaltman, Gerald (2003), "How customers think: Essential insights into the mind of the market," Harvard Business School Press, Boston, MA.
54. Zellner, Wendy (2005), "Wal-Mart: Your New Banker?," Business Week magazine, February 7th, 2005.

CHAPTER THIRTEEN

1. Fairfield Resources International Licensing Program. [Cited 20 April, 2005]. Available from http://www.frlicense.com/license.html.
2. Goldratt, Eliyahu M. and Cox, Jeff. The Goal. Great Barrington, MA: The North River Press, 1992.
3. Cahr, Darren S. and Morgan. Mary T. "Got Ownable Distinctiveness?" Brand Packaging, September/October 2001, 10.

CHAPTER FOURTEEN

1. Hrebiniak, Lawrence G. Making Strategy Work: Leading Effective Execution and Change, Wharton School Publishing (January 5, 2005).

A NOTE FROM THE PUBLISHER

Praveen Gupta, President of Accelper Consulting has been developing new business process designs for last five years based on his prior twenty years experience. The four new methods that he has developed are listed below:

1. Business Scorecard
2. Simplified Six Sigma Methodology and Corporate Sigma Level Method
3. 4P-Model of Process Management to Achieve Excellence
4. Breakthrough Innovation Methodology

These four tools constitute elements of his Strategy for Execution Map. Mr. Gupta believes that every corporation for profit has one fundamental strategy, "Achieve Sustained Profitable Growth." His strategy for execution map along with benchmarking provides the necessary capability to achieve the fundamental business strategy. Defining the fundamental strategy and developing tools to make it happen is not sufficient. Mr. Gupta understands the challenge to realize the fundamental business strategy. In order to support corporations achieve their fundamental strategy cost effectively, Mr. Gupta and his team offer additional tools and guidance. If you hit an impasse in realizing your corporate vision, and feel like talking to some one who has experienced a lot, you may like to ask Mr. Gupta for his guidance. His company, Accelper Consulting, can help you with the following capabilities, specifically in the field of innovation:

Understanding Your Needs: Experts from Accelper will visit your operations, meet with executives, and interview employees to understand the real issue using innovation diagnostic tools, and perform the gap analysis for identifying areas needing attention to accelerate innovation.

Planning for Innovation: The output from the gap analysis is then utilized for developing strategic plan suitable to your needs, and contingent upon your resources. The strategic planning for innovation

Mastering Business Innovation Certification: Accelper offers this unique training program to help your team learn basics of the innovation, and practice innovative thinking for developing solutions with breakthrough innovation.

Designing the Innovation Room for Breakthroughs: Corporations do need a dedicated space for innovation for perpetuating the environment for innovation. The room is designed to meet objectives of the innovation framework.

Establishing powerful measures of innovation: Accelper has developed expertise in defining practical measures of innovation. Mr. Gupta guides an executive team to establish measures of innovation, and their review for necessary actions.

CONTRIBUTORS'
BIOGRAPHY

Alex Goncalves is the Global Director of Quality Intelligence for Citigroup, Global Consumer Bank (New York, USA). He has been working in the field of quality management for almost 20 years and has experienced the evolution of quality over time. He has worked for different industries from construction to software development to financial services and consulting. He is an ASQ Fellow and a member of the Editorial Review Board of the Six Sigma Forum Magazine (ASQ/USA). He has published several articles about Quality Planning, Improvement and Control in the USA, Brazil and Argentina. He is currently the President of the Latin American Foundation for Quality (FLC/Argentina), and a Curator for the Brazilian Center for Quality, Safety and Productivity (QSP/Brazil). Alexis is also a recipient of the "Santos Dumont Medal" awarded by the Brazilian Air Force in recognition for his contribution to the development of Quality Management at the Aeronautics Institute (ILA/Brazil).

Hans Hansen teaches Organizational Behavior and Organization Dynamics courses at Victoria Management School, New Zealand, where he has encouraged creativity based on insightfulness attained through deep and critically understanding of material in light of reflection on our personal experiences. His research interests include Organizational Storytelling and Narrative Theory, Organizational Creativity, Organizational Culture, Cross-cultural Management, and Quantitative methods. Hans holds his Ph.D. from University of Kansas, and MBA from Baylor University. Currently, he teaches at VMS.

Dr. H. James Harrington serves as the Chief Executive Officer for the Harrington Institute and Harrington Middle East. Dr. Harrington is recognized as one of the world leaders in applying performance improvement methodologies to business processes. Dr. Harrington writes regular columns for the

Quality Digest Magazine. Dr. Harrington is a very prolific author, publishing hundreds of technical reports and magazine articles. He has authored 21 books, and 10 software packages.

In the book, *Tech Trending*, Dr. Harrington was referred to as "the quintessential tech trender." The New York Times referred to him as having a ". . . knack for synthesis and an open mind about packaging his knowledge and experience in new ways—characteristics that may matter more as prerequisites for new-economy success than technical wizardry . . ."

Rajeev Jain provides leadership in operational effectiveness and shareholder value improvement. He directs efforts in corporate development and international strategy for Hewitt Associates, a global human resources outsourcing and consulting firm. Instrumental in developing the organizational design and providing intellectual leadership for many performance enhancement initiatives at Hewitt, Mr. Jain is also a CPA (International), an MBA with specialization in strategy, and a Fellow of the Chartered Institute of Management Accountants and the Chartered Association of Certified Accountants. He lives in Hinsdale, Illinois.

Abhai Johri is Executive VP of Integral Consulting Services, Inc., where he manages deliveries of complex IT solutions. He uses creative approaches to Project Management and application design to deliver solutions that must exceed client's expectations. His interests include enterprise architecture, biometric-based identity solutions, Project Management, and creative teaching. He has published over 20 technical papers and holds five US Patents. Abhai has over 25 years of IT experience working for companies like IBM and Bell Labs. He received numerous awards for technical and management excellence, including IBM Chairman's Plateau Award and Bell Labs Project Excellence Award. He has BSEE from IIT Roorkee, India and MS from SMU, USA.

Laurie LaMantia helped start the new ventures organization in AT&T, and co-founded, IdeaVerse, a creativity and innovation center within Bell Laboratories. She has a MBA from Northwestern and is an adjunct professor at DePaul teaching

Creativity in Business, Leading from Within and *Entrepreneurship*. She is also the co-author of *Breakthrough Teams for Breakneck Times*, a guide for fostering collaboration.

Wayne H. Rothschild is the President of Innovation Management, LLC and President and Co-Founder of Neat-Oh! International, LLC. He is an inventor on more than 50 issued and pending patents, and an entrepreneur since age 10. He has held executive, management and technical positions with Kraft Foods, General Binding Corporation, WMS Gaming and LTV Aerospace and Defense. He has also held key leadership roles in successful start-up and turn-around situations. He has led numerous international due-diligence and technical evaluations with both major corporations and small businesses.

Mr. Rothschild has designed hundreds of products and manufacturing systems from nuts and bolts assembly to robotics and lasers. Many of these designs have earned him recognition and awards.

Mr. Rothschild holds an MBA from the Kellogg School of Management at Northwestern University and a Bachelor of Science in Mechanical Engineering from the University of Texas at Austin.

Justin Swindells is a patent lawyer who is also trained as an electrical engineer. Mr. Swindells practices patent law at the law firm of Jenkens & Gilchrist in Chicago, specializing in patent litigation and helping entrepreneurs obtain patents and trademarks from the U.S. Patent and Trademark Office. A judicial clerkship with a federal judge prior to entering private practice has given him a unique insight into the judicial decision-making process. And an upbringing spent on four different continents helped him develop a receptiveness to different viewpoints and solutions in achieving a client's goals. Mr. Swindells can be reached at jswindells@jenkens.com.

INDEX